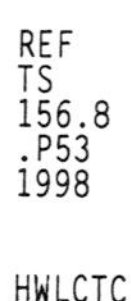

PROCESS CONTROL

A Primer For the Nonspecialist And The Newcomer 2nd Edition

By George Platt

67 Alexander Drive
P.O. Box 12277
Research Triangle Park, NC 27709

Printed in the United States of America.
10 9 8 7 6 5 4 3 2

ISBN 1-55617-633-3

Library of Congress Cataloging in Publication Data

Platt, George, 1920-
Process control: a primer for the nonspecialist and the newcomer
George Platt.
p. cm.
Bibliography: p.
Includes index.
1. Process control. I. Title.
TS156.8.P53 1988
670.42-dc19 88-15459 CIP

ISBN 1-55617-096-3

The Instrument Society of America wishes to acknowledge the cooperation of those manufacturers, suppliers, and publishers who granted permission to reproduce material herein. The Society regrets any omission of credit that may have occurred and will make such corrections in future editions.

Third Printing, 1993

Dedication

To Sarah -

my wife, friend, and helpmeet

TABLE OF CONTENTS

PREFACE

Who Should Read This Book?

Are you an executive with overall responsibility for the production, profitability, and safety of a process plant or other industrial plant? Have you moved up the company ladder, leaving behind much of your technical background? Or are you, perhaps, a whiz in sales or finance but never got into the technical side of things?

What do you do when your plant supervisor asks you to approve a project that requires an important addition or change to the plant instrument system? The project may involve obtaining and spending large sums of money, bearing the costs of downtime, complying with governmental pollution regulations, safety requirements, and so on. Does the plan seem technically practical? You could rely wholly on your technical people, but wouldn't a little knowledge of how instruments control a plant and of the technical lingo help you to make a more intelligent evaluation of the project?

This book surveys the field of process instrumentation and opens the door to understanding its principles and uses. It describes the elements of process control simply and with homey examples so that a non-technical person can understand its essential principles. It also describes the basics of some common types of instruments. The book discusses the importance of process control and its place in the overall plan for managing a company. It touches on engineering and managerial considerations for instrument systems.

Reading the book will not make you a competent instrument specialist, but you will no longer be a total stranger in a strange land. At least you will be able to ask more of the right questions. You may also find the subject very interesting.

This book may also be used in technical schools as an introduction to the ABCs of process control. The book may be useful to people - such as salesmen for machinery systems or electrical or safety engineers - whose concern with process instrumentation is incidental or passing.

The information in this book applies to a wide spectrum of industries, including:

- Architect-engineering.
- Industrial control.
- Process: food; metal refining and manufacturing; power generation; chemical; rubber; pharmaceutical; petroleum; glass; textile; paper; waterworks; sewage treatment; heating; ventilating, and air conditioning.

1

What Is Meant by a Process?

1-1 What Is a Process?

A *process* is any operation or sequence of operations involving a change in a substance being treated. Examples of processes include the following:

- A change of energy state, such as from hot to cold, or liquid to gas.
- A change of composition, as occurs in a chemical reaction or in mixing different materials.
- A change of dimension, as in grinding coal.

This book considers process in a broad sense because the applications of instrument principles are widespread. A process may be something as elaborate as deriving gasoline from the very complex mixture of chemicals in crude oil or something as simple as pumping water from one place to another. It may be heating a house to a desired temperature; pasteurizing milk; keeping an airplane at a constant speed, direction, and elevation; controlling a group of elevators in a building; operating an artificial heart; canning food; tracking a star by telescope; or tracking a ship by sonar or radar. For all these processes, certain universal principles of measurement and control apply, though the hardware and techniques may differ greatly. Also, the titles given to the people who supervise or direct the process may vary widely: plant operator, hospital nurse, airplane pilot, astronomer, radar technician, and many more.

This book focuses on instrumentation as it is used in industry, particularly the so-called process industries, which include those for making steel, treating foodstuffs, making chemicals, refining mineral ores, refining petroleum, and so on. The hardware mentioned in this book is typical of what these industries use.

Each process has a number of properties that may vary. Examples of properties are pressure, temperature, level, flow rate, acidity, color, quantity, viscosity, and many others. Each of these properties is known as a *process variable*. The changing values of the variables may be measured and sent to remote locations by means of signals. The measurements may be read, used for control, or stored.

1-2 What Are Process Instruments?

A *process instrument* is a device used directly or indirectly to perform one or more of the following three functions:

- *Measurement:* To measure is to determine the existence or magnitude of a variable (see Chapter 3). Measuring instruments include all devices used directly or indirectly for this purpose. Measurement systems may include auxiliary instruments to provide readings or alarms, to calculate derived values, or to perform other functions.

- *Control*: To control is to cause a process variable, known as the *controlled variable*, to be maintained at a specified value, or within specified limits, or to be altered in a specified manner. For example, room temperature may be controlled by a thermostat to remain constant. A controller requires and commands another device, which is usually a *final control element* (see the next function, "Manipulation"). An automatic controller may be looked at as an automatic brain that has no muscle. Different types of controllers may operate automatically or by manual adjustment, or either way (see Chapter 4).

- *Manipulation*: To manipulate is to cause a final control element to directly change a process variable in order to achieve control of another process variable. For example, the thermostat just mentioned operates a fuel valve, which manipulates the fuel flow. The final control element is the soldier, the muscle, that carries out the orders of the commander, the controller (see Section 2-1).

These three function categories are commonly referred to in the abbreviated form *measurement and control*. In this descriptive term, the manipulation function is absorbed into the control function because both are involved in the corrective action that controls the process. Nevertheless, it is important to be aware of the differences in function between control and manipulation.

The term *process instrumentation* covers the three function categories of instruments just described but, generally, it also includes related accessory instruments, instrument piping, and instrument wiring, if they are required. It excludes power supplies and other gadgetry, such as antifreeze devices, that merely enable the instrumentation to function.

1-3 Why Are Process Instruments Needed?

The history of mankind is tied intimately to the use of instruments. The measurement of land boundaries and distances; of the amount of periodic rise and fall of the Nile River and ocean tides; of time (by observing heavenly bodies); of the weight of cattle, precious metals, and the like—all these measurements in ancient times involved the use of primitive instruments. The age of automatic control began in 19th-century England with James Watt's invention of a governor to control the speed of a steam engine. This evolved into the modern universe of instruments, which have an almost limitless variety of function, construction, and purpose.

Now we are in the computer age in which computers are increasingly being hybridized with and paired with traditional instruments for measurement and control. The capabilities of instrument systems have been steadily increasing with resulting economic benefits for industry. These instrument capabilities include the following:

1-3-1 Range and Accuracy of Measurement

Instruments can measure dimensions over a wide range from the submicron to light-years. They can measure time intervals to less than a billionth of a second and weights from less than a millionth of a pound up to many tons—far beyond what any human can achieve. Extremely high accuracy is now commonly provided by automatic chemical analyzers, which can measure chemical composition to parts per billion. Only through instruments can we have reliably high-purity chemicals and metal alloys and know how good or bad our air and water are.

1-3-2 Process Operability

Some processes would not be practical or even possible if it were not for modern instrumentation. The whole field of nuclear power is a prime example of an industry that could not exist if it were not for the instruments that measure radioactivity. No one among us has any physical sensation when he has a dental X-ray or sits in his living room and receives natural radiation from the walls, but instruments exist that measure such things.

How would we measure the flow of catalyst for making gasoline in a fluidized-bed catalytic cracker if not for instruments? Without automatic instruments for measurement and control, could we get the exact mix of colors in the commercial cans of paint filled by the thousands in a paint factory? Skyscrapers would not be practical without elevators and their control systems to transport people and goods to the heights.

1-3-3 Process Quality

We take high quality for granted for so many products in our technological world. How many hundreds of millions of gallons of milk do we pasteurize properly every year for the health of our people? The pasteurizing process requires that the milk be held at an elevated temperature for some minimum time to kill harmful bacteria. At the same time, the temperature must not be too high and the heating time not too long or else too much nutritive value will be lost. Then the milk is checked for purity and stored at a controlled low temperature to maintain its quality. Maintaining this quality in large-scale production would not be possible without modern measuring and control instruments.

Suppose my car engine requires 88-octane gasoline and yours needs 92-octane. The oil refineries could choose to make only the higher-grade gasoline, which would be fine for your car and mine, but I would then have to pay the price for a better gasoline than my car needs. To satisfy both customers, the refineries produce both grades accurately through the use of sophisticated petroleum refining equipment and instrument systems. Thus, my car can have an economical fuel on which it runs well, and your "jazzy" car can have the more powerful fuel it needs.

1-3-4 Process Reliability and Safety

The results might be most unfortunate if the processes discussed in the previous section did not operate properly and safely. We might see frequent outbreaks of intestinal infections from improperly pasteurized milk and widespread dissatisfaction because automobiles had engine knock and got poor gasoline mileage.

Nuclear power plants are very complex. To enable all their nuclear systems to work well and safely, a staggering number of overlapping instrument systems and subsystems are installed to provide the high reliability required in the United States and other countries. All nuclear safety-related instrumentation is manufactured and installed to exceptional standards of conservatism and quality that far exceed what is normal for run-of-the-mill commercial instruments, which are themselves generally of high quality. (Further discussion of reliability can be found in Section 11-2.)

The following is an example of improving plant availability and reliability by using instruments to reduce the servicing frequency of process equipment. Feed water supplied to steam generators, especially large ones, should be of extremely high purity, close to that of distilled water. The purpose of such stringent standards is to minimize the corrosion of steam generator tubes and the buildup of scale, a mineral incrustation, inside the metal tubes. Corrosion eats away and thins the tube walls, leading to tube leakage. The scale hinders the transfer of heat from hot combustion gases to the water, thereby increasing the fuel usage by the steam generator; this incidentally lowers the operating efficiency and increases the operating cost. The scale also causes the metal tubes to operate hotter, thus weakening the metal and aggravating the corrosion. Continuous chemical analyzers are routinely used to keep the feed water adequately pure and thus protect the generator.

Another aspect of reliability is the fact that instruments do not have "people" problems; the instruments are more trustworthy.

1-3-5 Process Efficiency

The efficiency of a process may be very sensitive to how closely the process variables are controlled. Efficiency means how much of a desired output can be obtained from a given input to the process. For example, coal containing a certain amount of thermal energy is put into a steam generator and burned. A corresponding amount of steam theoretically would be generated if the plant operated at 100% efficiency. But no plant comes close to this ideal because of physical limitations and operating difficulties, including the weaknesses of the plant operators. If some of the human weaknesses can be eliminated or reduced by using instruments, the plant efficiency goes up. Higher efficiency means higher productivity and lower cost.

A high efficiency level requires not only that the process and the process equipment be designed well but also that the right instruments are used to make the process work the way it is supposed to.

1-3-6 Labor Cost

Many plant measurement or control functions that are performed by people can be performed at least as well by automatic instruments. The payback for the cost of instruments, including the cost of servicing the instruments from time to time, often justifies replacing people by instruments. The historical trend has been for process plants to have more and more instruments, both for their comparative economic advantage over people and for their superhuman capabilities.

2

How Instruments Work Together

2-1 The Sequence of Functions in a Loop

In Section 1-2, we stated that there are three general classes of instrument functions that are used in single instruments or in combinations of instruments to measure, control, and manipulate the process. All instruments perform one or more of these functions. A combination of instruments or functions that are interconnected to measure or control a process variable is known as a *loop*. Figure 2-1 shows a block diagram for a simple instrument loop to control the speed of an automobile; the control system is known as *cruise control*.

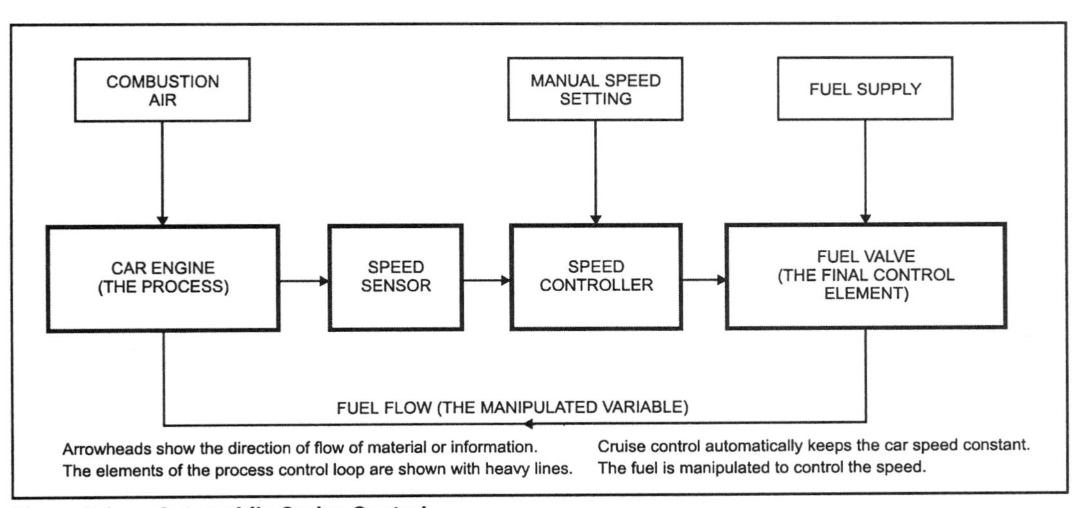

Figure 2-1. Automobile Cruise Control

The process is the car engine in operation. Our aim is for the car to automatically keep to a speed that we choose. Although there are technical variations of speed controllers, we will assume that the engine speed is an index of car speed, meaning that we will consider the car speed to be constant if the engine speed is constant no matter whether we are traveling uphill, downhill, or on the level.

To control the speed, the driver of the car manually sets a controller to maintain a speed of, for example, 45 miles per hour. The actual speed is continuously measured by a speed *sensor,* which sends a signal to inform the controller what the speed is. The controller then decides whether the current speed is too high or too low. If too low, the controller sends a signal to a fuel valve to open further to permit the fuel flow to the engine to increase; if too high, the controller commands the valve to be open less; if just right, nothing changes. In any case, the engine speeds up or slows down, if and as required. By continuously measuring, controlling, and manipulating, the cruise control system keeps the car moving at a constant speed.

Figure 2-1 shows how the sensor does nothing but measure and send out information. The sensor is also known as a *primary element*. The controller does nothing but study the process situation, comparing what is with what should be, and then sends out a command. The final control element does nothing but manipulate the fuel flow. (Here is where it is usually necessary to supply physical force to overcome process forces in order to give good control. In process work, the final control element sometimes applies great force in doing its job.) The engine speed is the controlled variable; fuel flow is the manipulated variable.

An example of another control loop, taken from industry, is shown in Figure 2-2. A constant volume of water is stored in a tank to provide a reserve for process equipment that withdraws water at fluctuating rates. Automatic level control is used to keep the tank level high and to avoid overflowing the tank. A sensor measures the water level and sends a signal to a transmitter that corresponds to the level. A *transmitter* is an instrument that passes on the information it receives from a sensor; the sensor may or may not be part of the transmitter. The transmitter then sends a measurement signal to a controller that directs the control valve, which, in turn, manipulates the water supply flow to the tank as required to keep the level constant. The controlled variable is water level; the manipulated variable is the flow of water to the storage tank.

The information sent out by the transmitter can be used by instruments other than the controller. Figure 2-2 shows how this same information can be used for other instruments, such as indicators, alarms, and so on. These instruments are not part of the control loop, though they are part of the measurement loop.

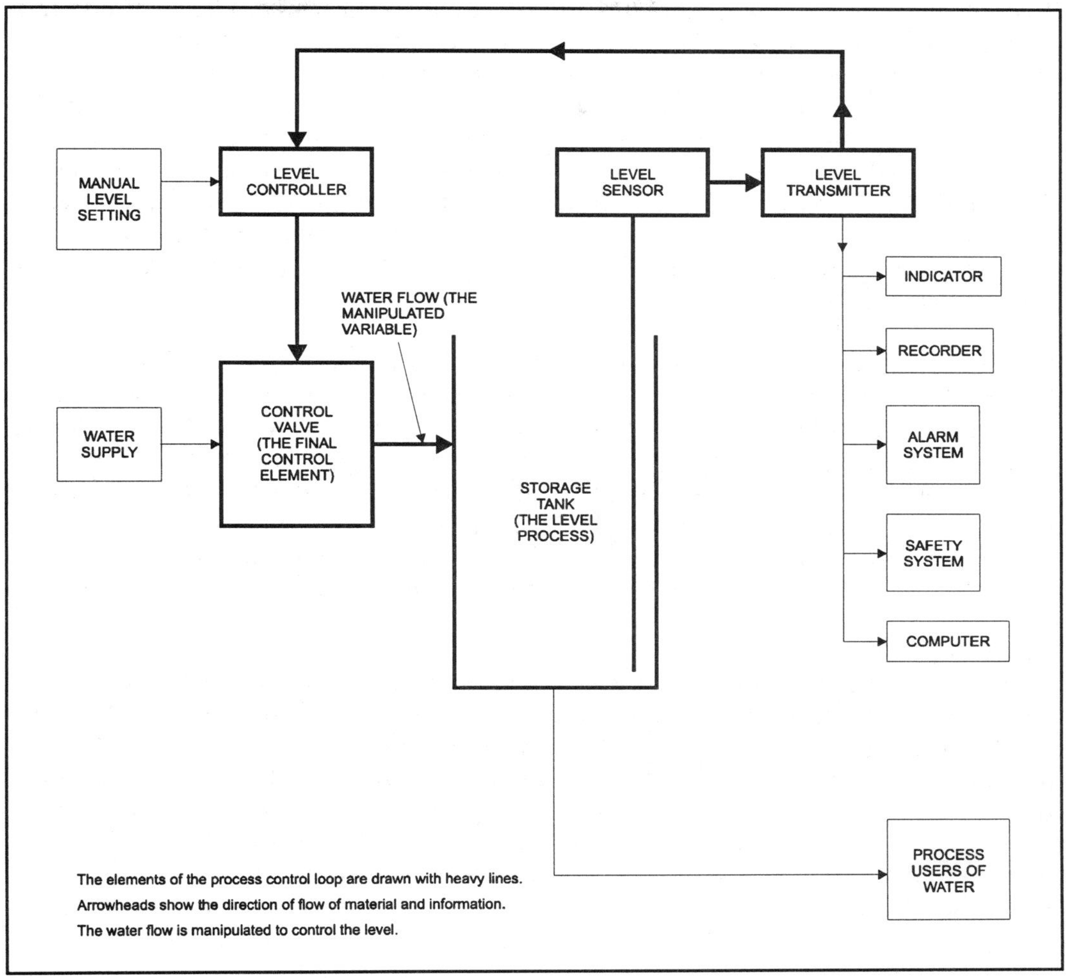

Figure 2-2. Control of Water Level in a Tank

2-2 Combining Functions in a Single Instrument

In order to measure or control a process variable, we may need only one instrument in some cases, but very often we need a group of instruments that work together, as discussed in Section 2-1. A household thermometer both senses and indicates the temperature; it is a complete measuring system, a complete loop. A common industrial loop uses two instruments to provide a temperature measurement that is to be read at a distant location. These loops are shown in Figure 2-3.

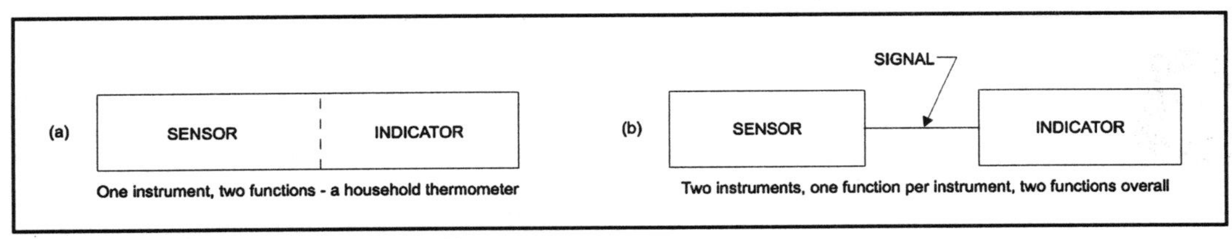

Figure 2-3. Two Loops to Indicate Temperature

Frequently, the level sensor and level transmitter shown in Figure 2-2 are separate instruments, but they can be joined in a single instrument. Thus, we often find many different combinations of sensing, transmitting, controlling, automatic switching, indicating, and other functions in single instruments, such as temperature-recording controllers, flow-indicating switches, and many others.

The individual functions are not always obvious when their hardware is in one physical assembly, yet the functions are there. By understanding the basic functions we can better understand how such assemblies work. As an example, consider a household iron, which is a single device that is a complete control loop in itself. The iron is a regulator that maintains a constant iron temperature that can be adjusted for ironing cotton, nylon, or other fabrics. Its use requires human guidance and human energy, but these are separate from its built-in function of controlling temperature automatically. It does this by measuring its own temperature, providing a corrective command if the temperature is not what the operator has set it to be, and closing or opening an electric switch. On low temperature the switch closes, electricity flows through an electrical heating element, the temperature rises. When the temperature gets too high, the switch opens, the flow of electricity stops, the temperature drops, and the on-off cycle starts again.

How Measurements Are Made

3-1 General Principles of Measurement

3-1-1 Direct versus Inferred Measurements

There are two general modes of measurement: *direct* and *inferred.*

- *Direct Measurement*. If we want to know whether it is windy outside, we can go outside and feel whether there is a wind. This is a *direct* measurement. In the case of instruments, there are flow meters, such as a household water meter or a gasoline dispenser at an automobile service station, that provide a direct reading of the fluid volume that has passed through.

- *Inferred Measurement*. We can determine whether there is a wind outside by going to a window and seeing whether our flag is waving, or whether the tree branches are swaying, or how rough the water is in the lake. These are *inferred* measurements; they depend on our sensing the effects of the wind instead of the wind itself. The Beaufort scale, developed by British Admiral Francis Beaufort in the days of sailing ships, is used by sailors to judge the strength of winds; it covers the range from calm to hurricane. Similarly, some flow meters determine fluid volume inferentially by sensing over a period of time the pressure effects that are caused by the flow. (For flow sensors, see Section 3-5.)

In general, an inferred measurement senses not the physical property that we want to know about but, instead, an effect created by that property.

3-1-2 Range and Span

Each measured variable and each measuring instrument has a selected range. *Range* is defined by a lower limit and an upper limit of variation or usefulness. The upper limit is known as full scale (F. S.). For example, suppose we manufacture a household thermometer for outdoor use where the temperature may vary between -20°F and +115°F. To provide some margin at the top and bottom, we extend the indicator scale on the thermometer so that it can measure -40 to +130°F. The *design temperature range* is then -20 to +115°F, and the *thermometer range* is -40 to +130°F.

Span is the algebraic difference, the numerical distance, between the lower and upper limits of a range. For our thermometer, the span is +130°F - (-40°F) = 170°F (see Figure 3-1). An automobile speedometer having a range of 0 to 110 miles per hour has a span of 110 miles per hour.

A range is always defined by two numbers, a span by one number.

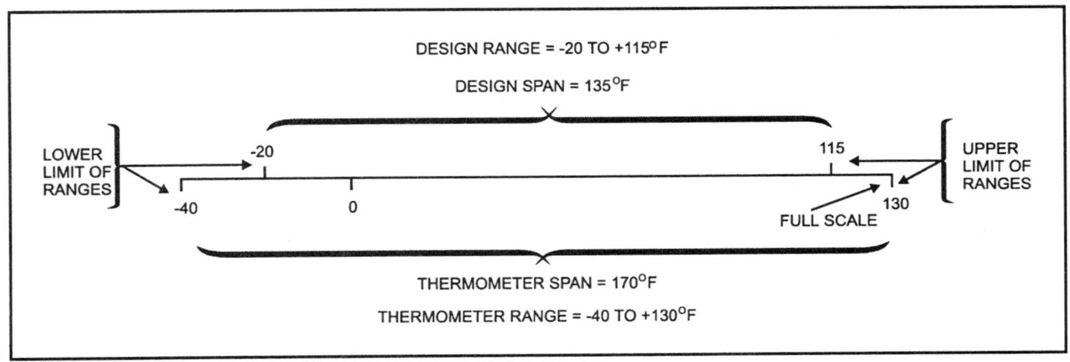

Figure 3-1. Instrument Range and Span

3-1-3 Static versus Dynamic Measurements

We have a pot of warm water on a table and we want to measure the water temperature. We put a thermometer in the water, and we wait a few minutes for the thermometer to be warmed by the water. Then we take a reading, which is easy to do because the water temperature has not changed.

Now we put the pot on a stove and turn the heat on to maximum. We find it difficult to read the temperature because the water temperature is continuously rising. Right now the temperature is 138°F, a moment later it is 139°F, then 140°F, and so on until the water boils. While the water temperature is changing, the thermometer responds but cannot keep up because it is chasing a moving target. When the water temperature stops changing, then the thermometer will be able to catch up.

When the pot of water was not being heated and the water temperature was constant, the measurement was made under *static* conditions, also known as *steady state* or *equilibrium* conditions. After the thermometer reached the initial water temperature, a good reading could be taken. But when the water was being heated and its temperature rose, the conditions were *dynamic*, also known as *unsteady state*. At a given moment when the water and thermometer temperatures were changing, their difference was the *dynamic error*. These two cases are shown in Figure 3-2.

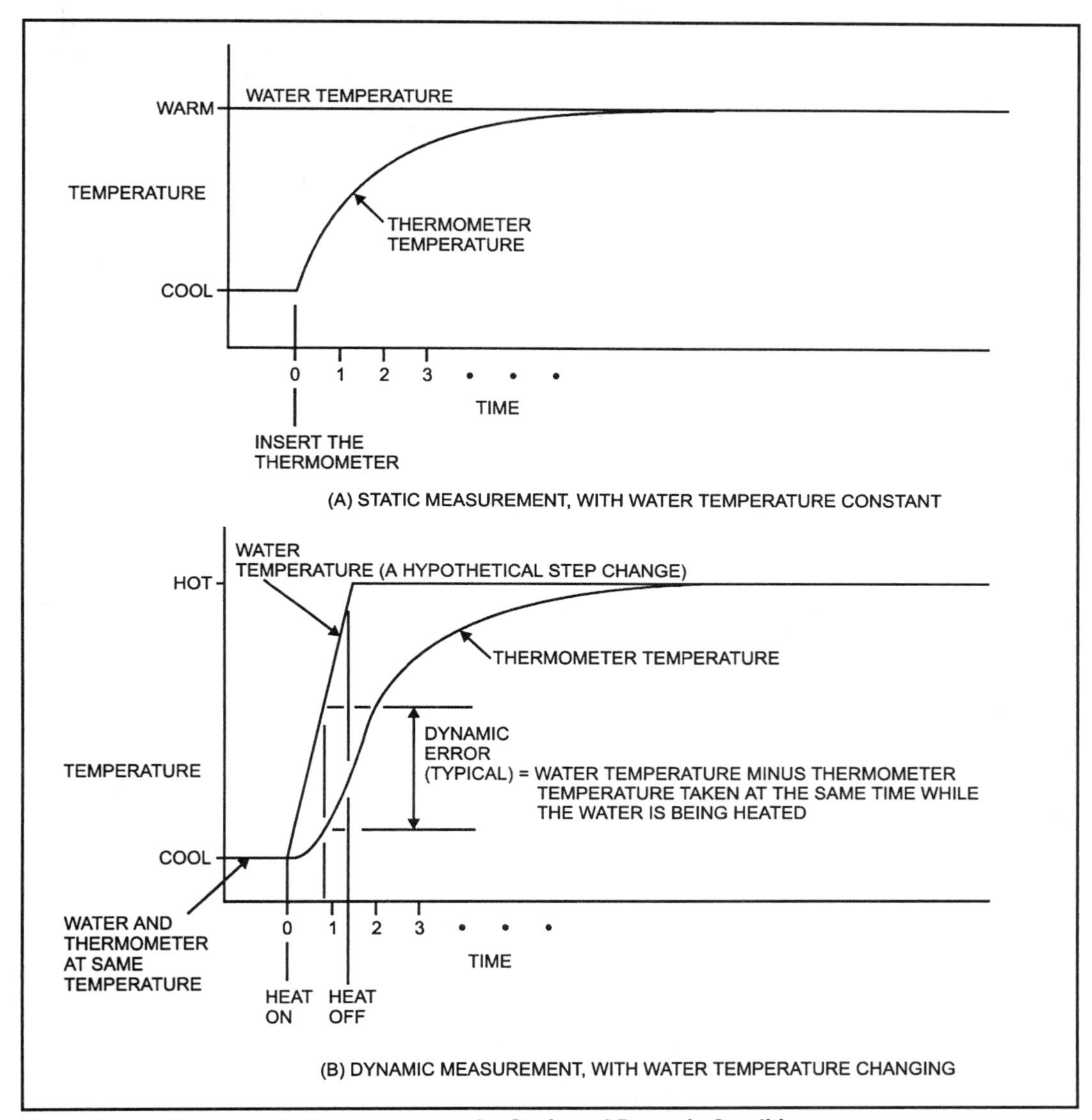

Figure 3-2. Temperature Measurement under Static and Dynamic Conditions

The term *dynamic* is relative. Some sensors are very quick to respond to changes and are said to have a fast response and a short response time; some have a slow response and a long response time. We might have little difficulty in tracking or keeping up with the water temperature changes if we used a fast-response sensor or if we heated the water slowly; the dynamic error would be small.

Figure 3-3 shows how an abrupt change of instrument input from its initial steady state to its final steady state causes a gradual change of instrument output. The abrupt change, which theoretically is done instantaneously, is known as a *step change*. A standard way of describing the response time for a simple system is by a *time constant*, which is based on a 63.2% change of output caused by a step change of input. For example, the time constant of a given instrument is stated to be six seconds, which means that, for a step change of input, the output will reach 63.2% of its ultimate change in six seconds.

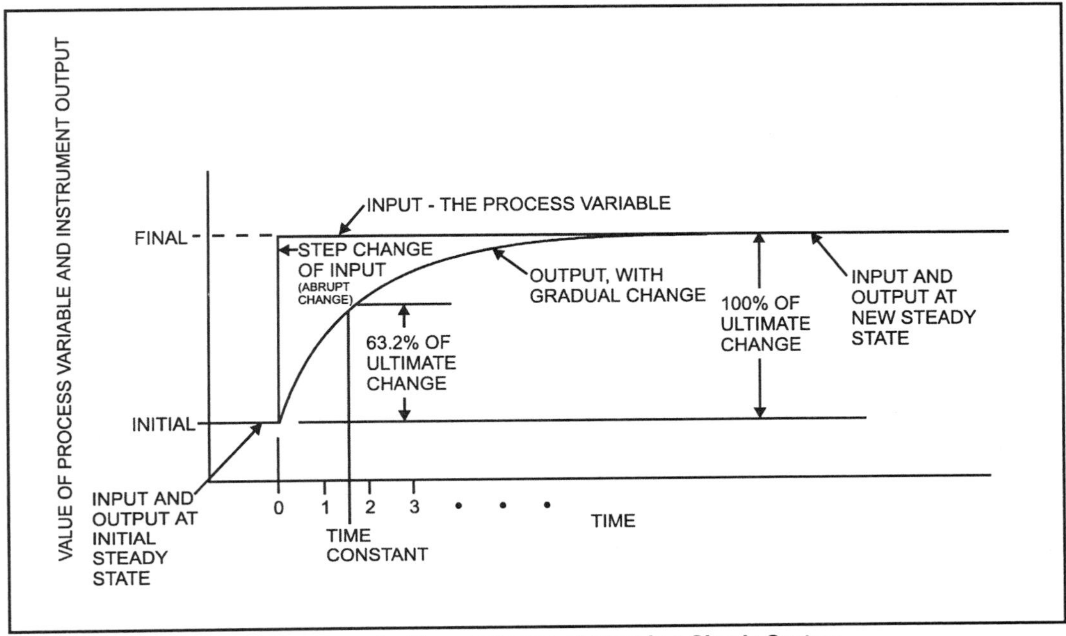

Figure 3-3. Instrument Response to Abrupt Change of Input in a Simple System

In comparing the response times for two different instruments, we have to be sure that the times are stated in the same way. When this is not done, it causes difficulties in evaluating the relative performance of different instruments being considered for purchase. Table 3-1 shows response time situations that may arise.

Table 3-1. Statements of Instrument Response Time

Instrument Response Time	Comment
Time constant is six seconds.	This statement is suitable for making direct comparison with other time constants, all of which relate to a 63.2% change.
Response time is nine seconds for 90% change.	We cannot say whether this response is faster or slower than that of a six-second time constant. The greater time of nine seconds relates to a change greater than 63.2%. Mathematical analysis is required.
Response time is four seconds.	Comparison with response statements for other instruments is not possible because the four seconds are not related to any specific amount of change.

3-1-4 Instrument Calibration

We speak of an instrument being calibrated. To *calibrate* a device means to determine its outputs that correspond to a series of inputs to the device. The data thus obtained are used to:

- Determine the locations at which scale graduations are to be placed.
- Adjust the instrument output to the desired value or values. For example, a temperature transmitter for measuring a range of 0 to 200°F requires an output signal range of 4 to 20 milliamperes (mA). The instrument is calibrated accordingly.
- Ascertain the error by comparing the actual output against what the output should be.

3-1-5 Truth in Measurement

All measurements are imperfect because the instruments that make the measurements are imperfect. There are limits to how finely any measurement can be made. The closest approach to the *True* value of a measured property—for example, the height of a table—is obtained only by using the best measuring tool available and the best measuring technique and taking the average of many measurements. Even the ultra-accurate measurements of the National Institute of Standards and Technology (NIST) (formerly the National Bureau of Standards) are imperfect.

The following are terms that describe how true measurements are and how much faith we can place in them.

3-1-5-1 Accuracy

The room we are in happens to have an actual temperature of 72.128641 ... °F even though we do not know this. We look at the room thermometer, which shows a temperature of a hair over 72°F, and we announce that the temperature is 72 degrees. We cannot read the exact temperature no matter how careful we are. We could use a laboratory thermometer and read the temperature as 72.13°F or even 72.129°F, but there would still be a residual error. We never reach the absolute truth; we only approach it more or less.

A measurement error equals the reading, the apparent value, minus the true or ideal value. In practical terms, let us say that our actual room temperature is what a very accurate thermometer says it is: 72.129°F. The error is then 72°F minus 72.129°F, or -0.129°F. A negative error means that the reading is less than the true value; a positive error means that the reading is higher than the true value. The error in a given case may or may not be important. For us, sitting in the room, who cares if we think the temperature is 72°F instead of 72.129°F?

A peculiarity of engineering terminology is that an error, which is the *inaccuracy,* is usually called *accuracy.* This "accuracy" is -0.129/72 = –0.0018 of reading = –0.18% of reading.

The way that comparisons are stated can cause great misunderstanding when applied to accuracy or other things. For example, you put some money into a bank account, and you receive 6% interest. Later you switch to a different kind of account and you receive 8% interest. You tell me that your interest rate has increased by 2%. I say it increased by 33 1/3%. Can we both be right?

The problem here is that each of us is using a different and unstated base of reference. If you will say that the rate has increased by 2% of the balance in your account, and if I will say that the rate has increased by one-third of what it was previously, then we can understand each other and agree that both our statements are correct.

The accuracy of an instrument can also be stated in several different ways. Assume that an instrument has an indicator range of 50 to 850°C. Its accuracy may be stated as shown in Table 3-2. The words *of full scale, of span, of range,* and *of reading* are sometimes omitted through oversight or in the belief that the meaning of *percentage* is clear in context. If doubt exists, the appropriate words should be used to avoid misunderstanding.

High accuracy, meaning low error, is preferred to low accuracy. Whether high accuracy is worth its extra cost and other complications is a question that must be answered for each application.

Table 3-2. Statements of Instrument Accuracy

Accuracy Statement	Remarks
±0.1% of full scale (F.S.)	±0.1% × 850°C = ±0.85°C at any point in the range. Acceptable statement. This leads to large errors for readings that are low in the range, thus: For a reading of 500°C, accuracy = ± 0.85°C/500°C=±0.2%. For a reading of 100°C, accuracy = ±0.85°C/100°C= ±0.9%.
±0.1% of reading	For a reading of 500°C, accuracy = ±0.1% × 500°C= ±0.50×C. For a reading of 100°C, accuracy = ±0.1% × 100°C= ± 0.10°C. Acceptable statements.
±0.1% of span	±0.1% × 800°C = ±0.80°C. Acceptable statement.
±0.1% of range	Improper statement. A range is defined by two numbers.
±0.1%	Incomplete statement. A reference value, such as *full scale,* is required.
±0.85°C over the full range	Acceptable statement.
±0.25°C at 25% of span	Acceptable statement.
±0.25°C at 250°C	Acceptable statement.
±0.25°C	Incomplete statement. A reference value, such as *full scale,* is required.

The accuracy of each of the instruments in a loop is typically poorer than the accuracy of the loop. This is because the loop accuracy includes the degrading effect of all the individual up and down inaccuracies. (This is discussed further in Section 11-1.)

3-1-5-2 Accuracy versus Precision

The words *accuracy* and *precision* do not have the same meaning. Accuracy is the closeness by which the apparent value of a variable matches the true value or an accepted standard value of the variable. Or simply, it is a comparison between a reading and the truth.

On the other hand, precision is the closeness to each other of successive apparent values of a variable. Or simply, it is a comparison between different readings of the same thing. It is a measure of the scatter of supposedly equal values. Figure 3-4(a) illustrates the difference between accuracy and precision.

Consider an unadjusted rifle that shoots far to the right of the target but whose shots are very closely spaced. The rifle has low accuracy with high precision, analogous to the illustration of Figure 3-4(a), line (iii). After the rifle sight is adjusted, both the accuracy and the precision are high, as in Figure 3-4(a), line (i).

3-1-5-3 Repeatability versus Reproducibility

Assume we change an input to an instrument over its full range by steps up and down and we measure the corresponding output values. If we do this several times under the same conditions, we end up with the information shown in Figure 3-4(b). Because of various imperfections of the instrument, the several lines of output versus input in the up direction do not coincide with each other, the several lines in the down direction do not coincide with each other, and the up lines do not coincide with the down lines. If the instrument operated perfectly, there would be only one line instead of six lines.

For a given input, the span of extreme upward values of output is the *upward repeatability,* and the span of downward values is the *downward repeatability.* For that input, the span of the extreme values of up and down outputs is the *reproducibility.* These terms are illustrated in Figure 3-4(b). The lack of coincidence of lines that would ideally coincide reveals the nonrepeatability and the nonreproducibility; nevertheless, these are usually expressed as *repeatability* and *reproducibility,* respectively, in percentage of span.

Reproducibility is a more useful word than *repeatability* because it includes the effects of repeatability as well as other inaccuracies—drift over a specific period of time, hysteresis, and dead band—for the instrument in question. A reproducibility requirement is not usually put into a specification; it is covered under the blanket term *accuracy,* which is frequently specified. However, there are applications for which good accuracy of measurement or control are not required but that do require good precision, that is, good reproducibility and predictable performance.

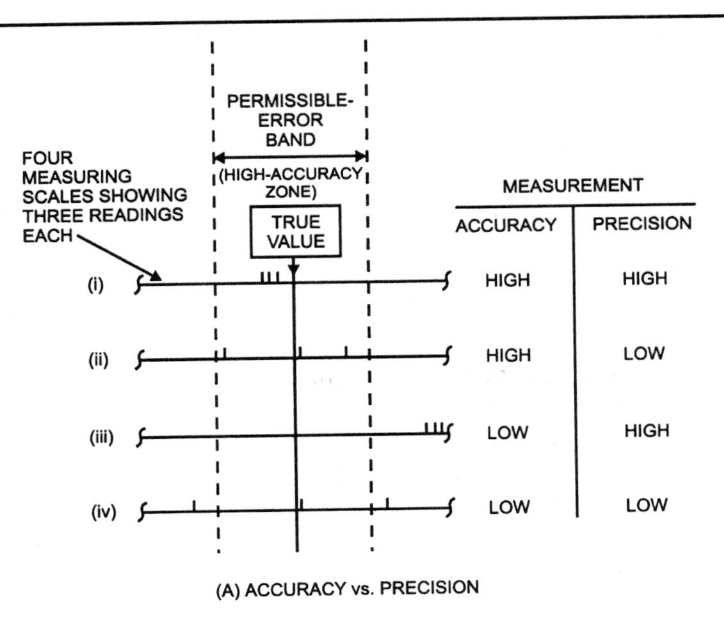

(A) ACCURACY vs. PRECISION

The goal is to measure the true value of a process variable. Any measurement within the permissible-error band is considered to be of high accuracy, outside the band to be of low accuracy. A small scatter of readings shows high precision; a large scatter, low precision.

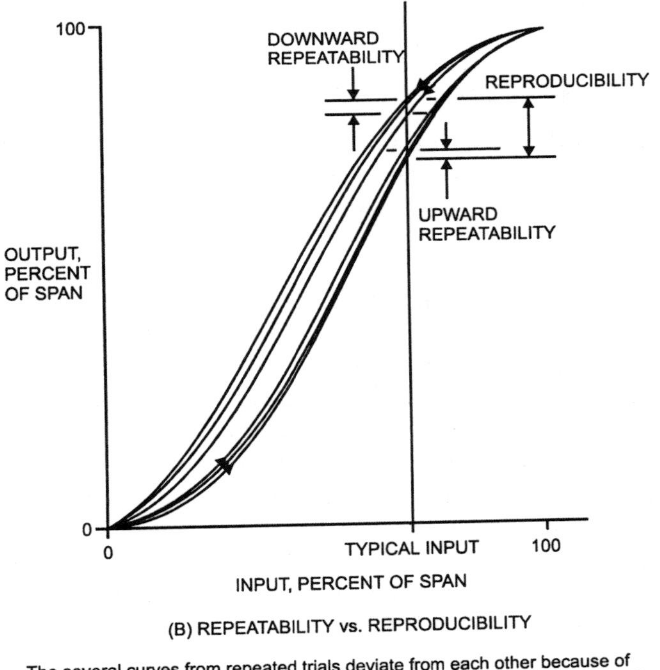

(B) REPEATABILITY vs. REPRODUCIBILITY

The several curves from repeated trials deviate from each other because of instrument inaccuracy. The ideal would be a single straight line from 0 to 100 percent of the input and output spans.

Figure 3-4. Truth in Measurement

3-2 How Information Is Conveyed

3-2-1 The Information Chain

Input and *output* are essential words for discussing process instruments.

- *Input* is any information or signal that is supplied to an instrument or other device. The information may come from contact with a process, as when sensing temperature, flow, liquid level, chemical composition, or any other process variable. Also, the information may come in the form of an electric, pneumatic, hydraulic, fiber-optic, thermal, or other signal. The input to the receiving instrument may be used (a) directly, without modification, if appropriate, or (b) indirectly, by modifying the information if and as necessary and then passing it on as the output. This output is, in turn, the input to still another instrument.

- *Output* is the information or signal derived from an instrument or other device. The output may be in one of the following forms as well as others:

1. A signal to one or more other instruments.
2. An indicator reading.
3. An on-off light or an alarm bell to show process status.
4. A record such as a graph, printed chart, or magnetic tape.
5. A message on a computer screen.
6. The process flow manipulated by a control valve.

The engineering words *input* and *output* are applied not only to individual instruments but also to instrument systems. (They are even used in nonengineering situations, for example, "We need more input to solve our problem.")

For the instrument loop illustrated in Figure 2-2, Figure 3-5 shows how the output from one instrument is the input to the next instrument. The first input to the loop is information about the process water level. The level sensor, followed by the level transmitter, transform the level into first one signal then another as the level changes, and the transmitter informs the various receiving instruments accordingly. The level indicator, for example, has no output signal but provides a reading of the level for the plant operator. The level controller has a manual adjustment for the desired water level; this is a second input. The controller has an output signal that is the input to the control valve. The control valve is the final control element whose output is the flow of process water to the tank; as the valve input signal changes, so does the water flow change.

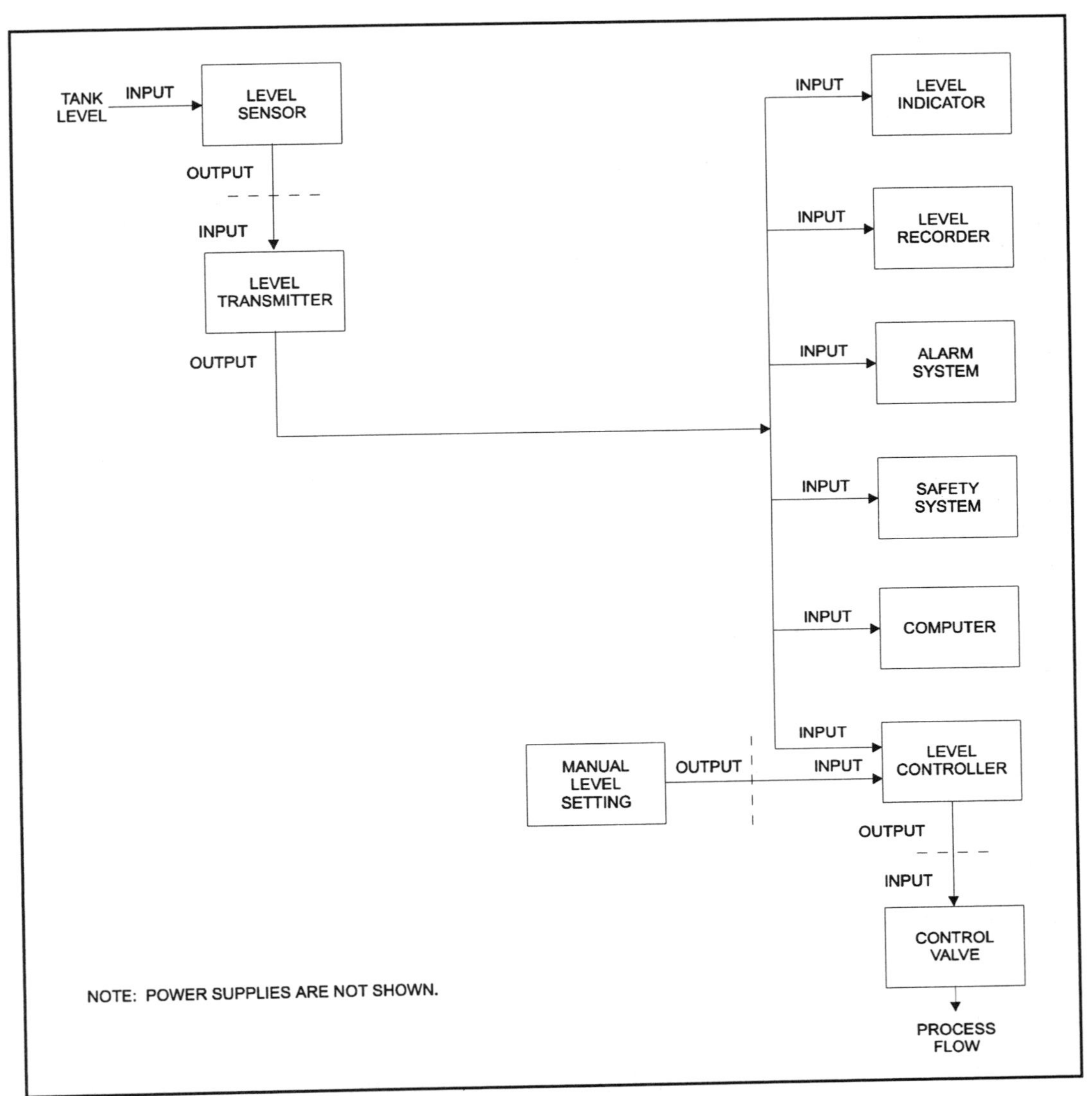

Figure 3-5. Information Chain for an Instrument Loop

Many, if not most, instruments require a power supply in order to function. The supply is usually electrical or pneumatic. An unmanipulated power supply, as such, is not an input to instruments, even though the power may be indispensable for their functioning; the power supply is simply an energizing medium that contains no information. For example, to watch a television program we turn our set on. This does us no good at all if the selected channel is not broadcasting information.

3-2-2 Types of Instrument Signals

A signal is a variable that is made to represent a different variable in a specific relationship. For example, a process variable, such as liquid level in a tank, may be represented by an air pressure, an electric current or voltage, or a property of some other signal medium. Similarly, a signal may be altered through one or more instruments to provide a desired result, as is done by the level controller in causing the control valve to manipulate the flow of water.

There are three basic types of signals: *binary*, often called on-off; *analog*; and *digital* or *pulse*. They are described in the following three sections.

3-2-2-1 Binary Signals

A binary signal is the simplest type of signal. It has two discrete values, *On* or *Off*, meaning *Yes* or *No*, or *True* or *False*. The values can be symbolized as 1 or 0, respectively. (*Discrete* means consisting of *unconnected and individual parts*.) An instrument output binary signal changes from one value to the other according to whether the input is less than or greater than some given value. But depending on the application, the input may be either binary, analog, or digital (see Sections 3-2-2-2 and 3-2-2-3).

An automatic home thermostat is an example of a device providing a binary signal. Assume that the thermostat is set to maintain the room temperature at 70°F but the actual temperature is 65°F. The thermostat is on, an *On* signal is being sent to the final control element, which is either a gas valve or an oil pump, and fuel is flowing to the home furnace; the room gets warmer. When the temperature rises to 70°, the thermostat goes off, the signal goes off, and the fuel flow stops. The room cools and the cycle repeats, with the thermostat, in effect, continuously asking: Is the temperature below 70°F? If *Yes*, the thermostat commands that the heat be turned on; if *No*, it commands that the heat be turned off. The thermostat output signal is always either *On* or *Off*, with nothing in between.

A binary signal is sometimes referred to as a *digital signal* or a *discrete digital signal*. However, a binary signal at any moment states a complete piece of information concerning a condition, whether *On* or *Off*. The signal state may last only momentarily or indefinitely. It does not have the periodic and repetitious nature of a true digital signal. The output of the level controller in Figure 3-5 may be binary.

Many binary instruments are *direct-acting*, which means that an input that is lower than its desired value causes a low output; an input above its desired value causes a high output. Also common industrially are *reverse-acting* instruments, for which a low input causes a high output and vice versa.

3-2-2-2 Analog Signals

You may have seen a jewelry store display of a watch whose face has no dial graduations and represents a surrender of function to fashion, yet it is readable ... somewhat. Despite the lack of markings on the dial, if the watch has the customary twelve-hour movement, we can see the approximate time from the position of the hands. The movement of the hands is analogous to the passage of time because the hands move as the time advances. Hence, such a timepiece is called an *analog watch*. For instrument work, an *analog device* may be defined as something that moves or changes without a break as something else moves or changes without a break.

Examples of everyday analog instruments are a household thermometer and an automobile speedometer, which give readings that are analogs of the changing temperature and speed, respectively. Such instruments, whose design was not ruled by high fashion, are more practical than the aforementioned watch and have scale graduations to permit more accurate readings. Instrument signals are also often of the analog type, whose value is analogous to the value of the information that went into the instrument. The automatic adjusting of the analog output of an instrument as the input changes is known as *modulation*.

For most analog instruments, the output varies in a one-to-one relationship with the input; for every input value, there is one specific output value. For certain controllers, the output varies as the input varies but also varies with time in a definite way; the output of these controllers can have different values at different times even if the input does not change (see Sections 4-2-3 and 4-2-4).

Input and output ranges are generally stated using percentages that relate to the span of each range (see Section 3-1). However, the scales for a specific instrument are commonly stated in physical terms; for example, a flow input may be 0 to 300 standard cubic feet per minute (scfm) and the signal output may be 4 to 20 mA dc.

3-2-2-3 Digital Signals

A digital signal has discrete elements, typically a stream of pulses whose height, frequency, or shape are analogous to the way the signal device input changes value. Though the signal has a pulse form, the information it carries may end up in one of the following forms: binary, using zeros and ones; analog, a continuous stream; alphabetical; numerical; or alphanumeric, which is a combination of letters and numbers.

Signal converters can change pulse-form digital signals to the other forms and vice versa. Thus, intelligible messages may be sent. Examples include instruments that send out digital signals to represent a pressure, temperature, or other variables or the common digital wristwatch which counts electrical pulses and provides a time reading using the ten digits of the decimal numbering system.

3-2-3 SIGNAL RANGES

Binary signals, by definition, have only two values, a lower one and an upper one. Analog signals have an unbroken series of values from the top to the bottom of their range. Common instrument signals are shown in Table 3-3. Instrument Society of America (ISA) Standards S7.0.01, *Quality Standard for Instrument Air*, and S50.1, *Compatibility of Analog Signals for Electronic Industrial Process Instruments*, define the requirements for pneumatic and electronic signals, respectively.

There are also electromagnetic signals: optical (using light rays), infrared (using heat rays), and radioactive (used for measurements), which are not listed in Table 3-3. Also, some measurements are made by acoustic signals.

The reason why some analog signal ranges—for example, 3 to 15 pounds per square inch gage (psig) or 4 to 20 mA—do not start at zero is to improve the accuracy at the low end of the range. The 3-psig and the 4-mA values are known as *live zeros*.

Table 3-3. Nominal Signal Ranges

Type	Medium	Values
Binary (on-off)	*Electricity:*	
	Alternating current (ac)	0 or 120 volts (V)
	Direct current (dc)	0 or 24, 48, or 125 volts
	Pneumatic	0 or 25, 35, 100 pounds per square inch gage (psig)
		(170, 240, 700 kilopascals [kPa])
	hydraulic	0 or 3000 psig (20,000 kilopascals)
Analog (modulating)	*Electricity:*	
	Direct current	-10 to +10 volts, 1 to 5 volts*, 4 to 20 mA*, or 10 to 50 mA
	Pneumatic	3 to 15 psig*, 6 to 30 psig (20 to 100 kilopascals, 40 to 200 kilopascals)
	hydraulic	0 to 3000 psig (0 to 20,000 kilopascals)
Digital, discrete (with open-collector transistor):		Transistor-to-transistor logic (TTL) 0 or 5 volts*; 0 or 12 volts*
Digital, serial data transfer:		RS 232*; RS 422*; RS 485*
*The noted ranges are standard.		

3-2-4 Converting Signals

Figure 3-5 shows the output of the level transmitter sending information to six receiving instruments. Assume that the transmitter output signal is compatible with the input requirement for all the receivers except the alarm system and the computer. Figure 3-6 shows how this problem is resolved by adding signal converters. Converters are available for all common combinations of signal ranges and can be developed for all others.

The identification numbers used in Figure 3-6—LV-1, LE-1, and the others—are the individual numbers for the instruments shown.

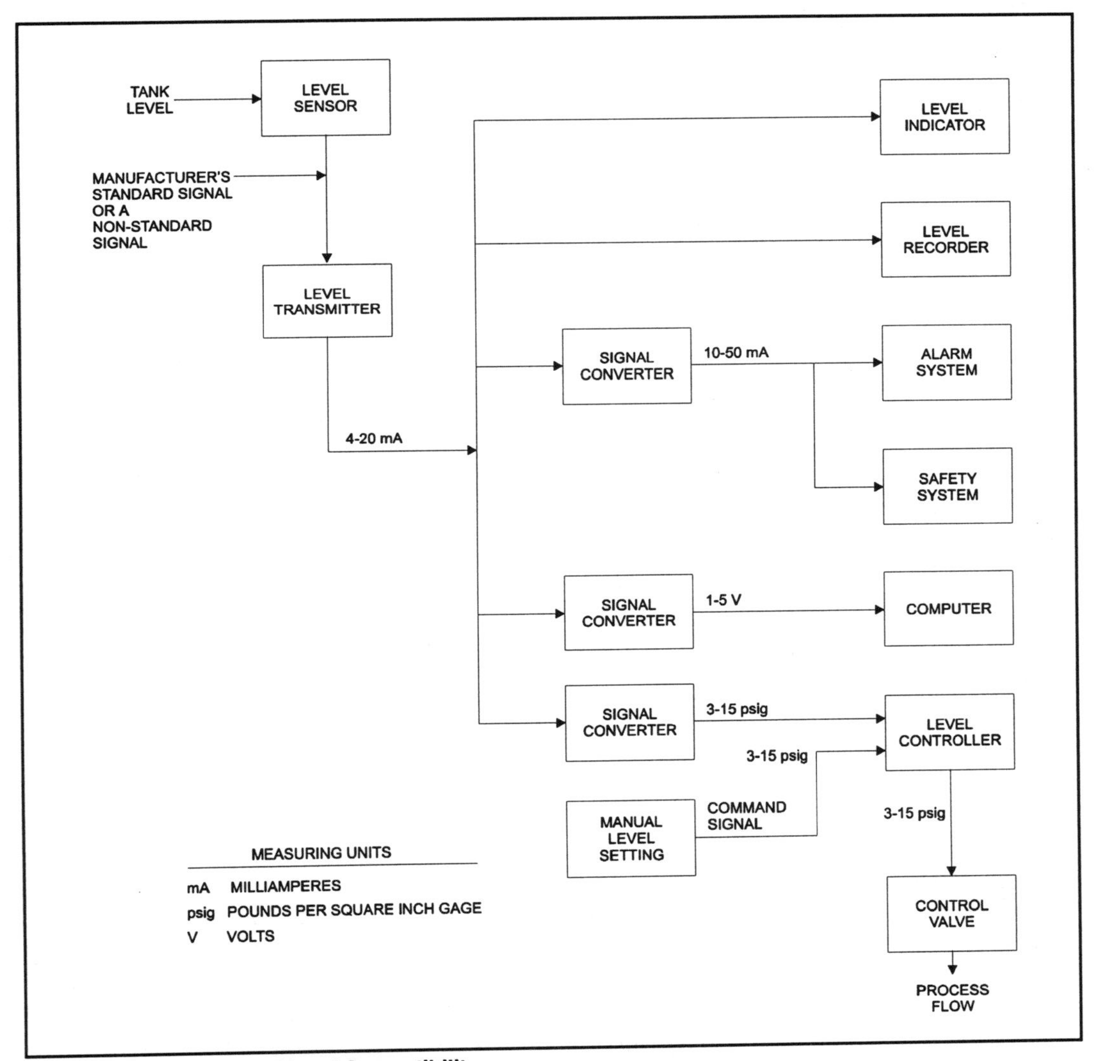

Figure 3-6. Providing Signal Compatibility

3-2-5 Sharing a Signal Channel

Often a number of signals have to travel a long distance or undergo a similar modification of some sort. *Multiplexing* is a system that enables such signals to simultaneously transmit two or many messages in either or both directions over the same path. This makes data highways possible for use in distributed control systems. A *data highway* carries information to and from many instruments by means of either twisted pairs of wire, coaxial cable (coax), or fiber optics of plastic filaments or glass filaments that transmit light. Coaxial cable has a central conductor and an outer shell.

Timeshare multiplexing uses signals that take turns in sharing a single processing instrument or a single communication channel. The switching is termed *scanning*. Figure 3-7 illustrates the basic concept of multiplexing.

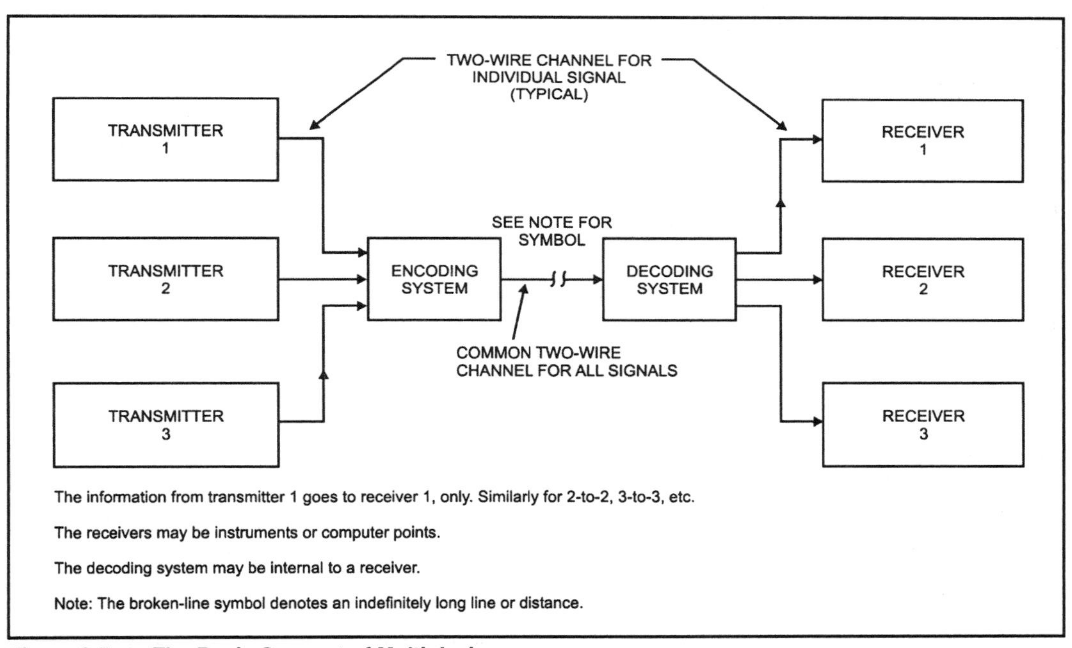

Figure 3-7. The Basic Concept of Multiplexing

3-2-6 How Instruments Are Identified and Symbolized

Figure 3-6 shows a group of instruments for a single instrument loop. A plant may have many dozens of instrument loops; a nuclear power plant may have up to 40,000 or more individual instruments. An efficient scheme is needed to enable each instrument of the loop to be identified for use in designing the loops, buying the instruments, warehousing, installing, testing, servicing, writing plant operating instructions, and so on. Also, an efficient scheme is required to symbolize the instruments for engineering and construction drawings. ISA

Standards S5.1, *Instrumentation Symbols and Identification,* and S5.3, *Graphic Symbols for Distributed Control/Shared Display Instrumentation, Logic and Computer Systems,* address these needs.

Instrument number or *tag number* is the identification number assigned to an instrument. Figure 3-8 shows the ISA method applied to the system shown in Figure 3-6.

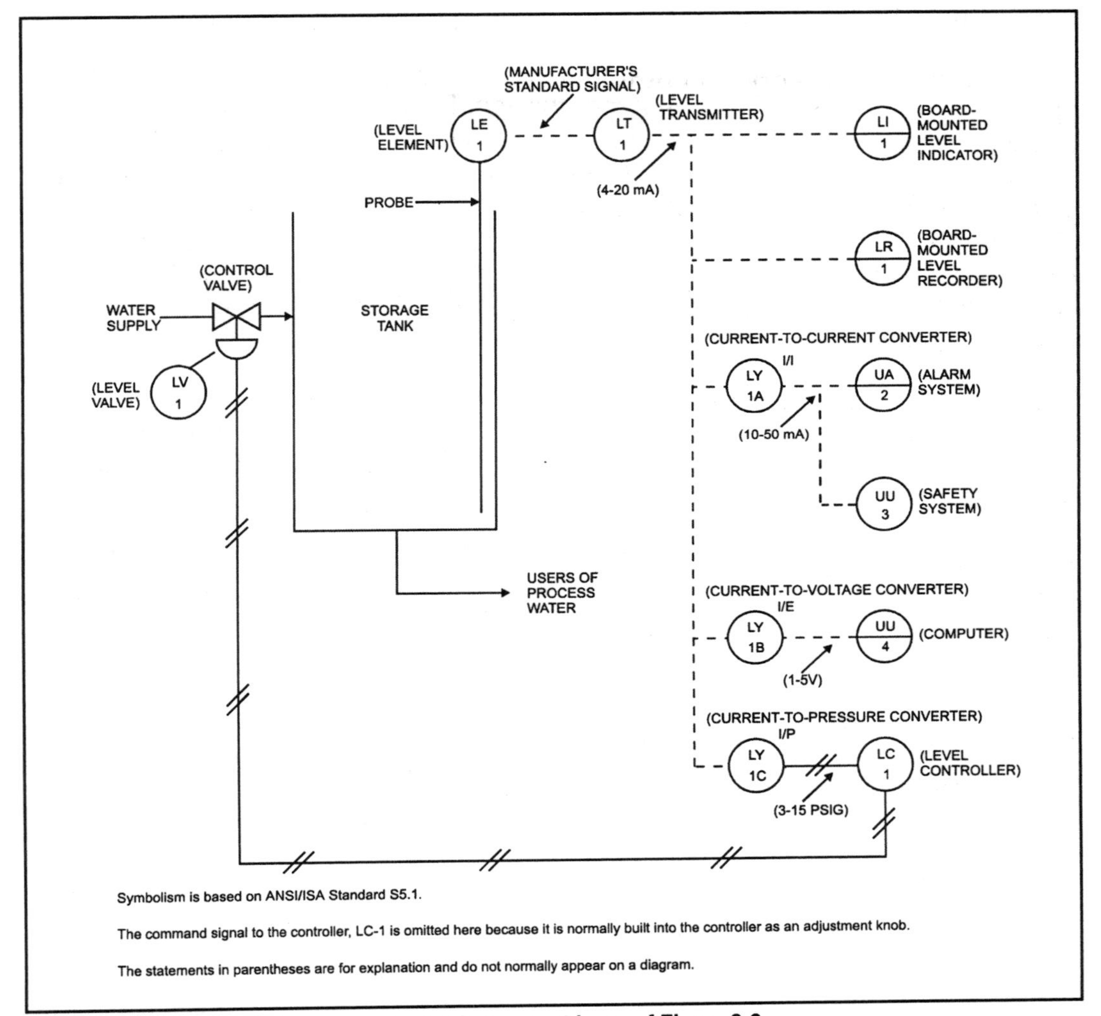

Figure 3-8. Standard Symbolism for the Instrument Loop of Figure 3-6

3-3 How Information Is Displayed and Retained

Much of the information provided by instruments has to end up in some sort of display that can be read by people. An example is statements of current process conditions that an operator has to keep his or her eye on to see whether the plant is running normally. If the conditions are not normal then the operator has to decide what to do about it. This is similar to the driver of an automobile checking the dashboard instruments to see that the car is performing properly.

Past information—historical information, as it is called—may be important. The "past" may be five minutes or five months ago. Before the operator decides what corrective action he or she should take, he or she may want to look back in time and see how the process behaved: its direction of change—toward normality or away from it; the speed of change; or what caused a failure or accident to happen. Past information is needed also to bill a customer or pay for fuel or raw material. Did pollutant emissions for the past week exceed government regulations? How efficiently has the plant operated over the past month? How excessive was the operating pressure and for how long?

Even future information can be presented, though this is uncommon. Future information makes use of predictive instruments that determine how a process system will behave at some time in the future based on what is known about the present state and the characteristics of the system.

Information may be classed as either *nonpermanent* or *permanent*. The nonpermanent information is for immediate use. The permanent information may be for immediate use but is also retained for later use.

3-3-1 Presenting Nonpermanent Information

Nonpermanent information may be presented using glasses, indicators, lights, annunciators, and video screens. These are discussed in the following sections.

3-3-1-1 Glasses

Glasses are the simplest form of display. They consist of a piece of transparent glass or plastic, usually encased in a metal housing, that is connected to the process and shows what is going on inside.

Glasses may not have a measuring scale; if they do, they are called *indicators*. The most common glass instrument is the *level glass, level gage*, or *gage glass*, which is connected to a vessel and shows the liquid level (see Section 3-6-2). Another example is the *flow glass* or *sight-flow indicator*, which is connected to a pipeline and has a vane, a paddlewheel, or a drip arrangement inside to show flow. Depending on the flow rate, the vane swings open more or less, the paddlewheel turns faster or slower, or the drip stream is stronger or weaker.

3-3-1-2 Indicators

An indicator has a graduated scale and a pointer or index mark, as typified by a common liquid-in-glass thermometer and most automobile speedometers. The indicator shows the value of a process variable. Examples of analog scales are shown in Figure 3-9. Indicators may be digital and display an exact reading, such as "128.1°F". An automobile speedometer, which shows the speed of travel, is almost always analog. Its companion odometer, which shows the distance traveled, is digital.

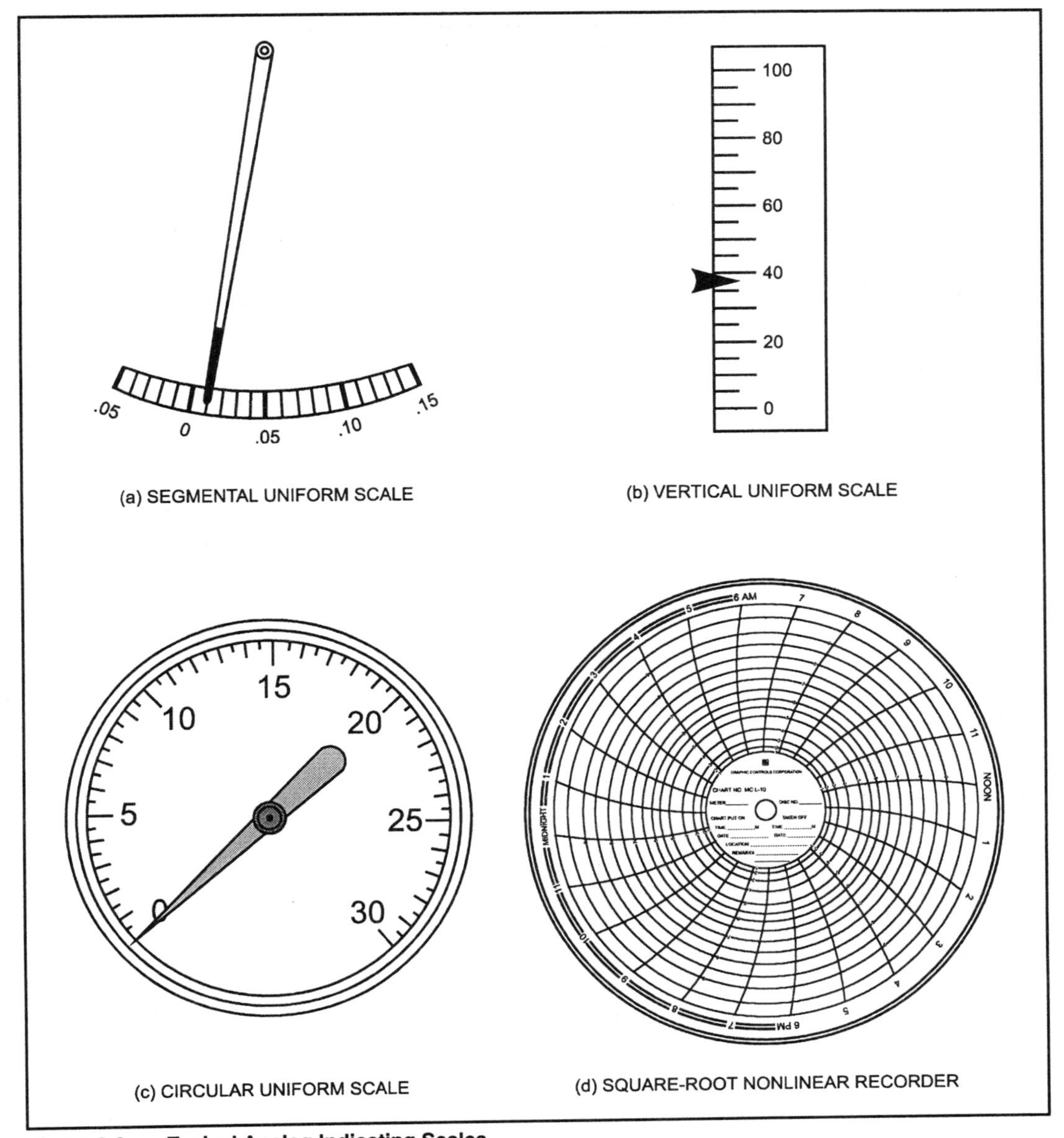

Figure 3-9. Typical Analog Indicating Scales

3-3-1-3 Lights

Small lamps with colored translucent covers are widely used to convey binary information. A lamp is either *On* or *Off*, sometimes *Bright* or *Dim*. The information shown may report that the status is normal, for example, a valve is open or a pump is running. A light may show that a particular operating sequence is under way or is completed; some automatic dishwashers and clothes driers have such lamp displays. A lamp that signals which one of a number of normal conditions exists is called a *pilot light*, though some refer to it as an *indicating light* or a *monitor light*.

Alarm lights are used to warn of an abnormal situation that may lead or has already led to trouble or danger. Alarms are placed in automobiles for abnormal and potentially hazardous conditions such as low oil pressure, high oil temperature, seat belts not fastened, or doors open.

In a process plant, two separate types of alarms are commonly provided. The first alarm is actuated during an abnormal state but before harm has been done. It is a *first-stage alarm* or *preliminary alarm*, which is often shortened to *pre-alarm*. The first-stage alarm notifies the operator, who should then take action to avoid a potential danger. If and when the condition gets worse, the operator gets a *second-stage alarm*, or simply an *alarm* if a pre-alarm came first. The system that actuated the second-stage alarm may or may not cause an automatic shutdown of a part or all of the process, but, in any event, the operator has to go through an appropriate emergency procedure.

Alarm lights and pilot lights are effective only if an operator sees them. They do not necessarily command instant attention. To help the operator, an alarm light may flash to signal trouble. *Flashing* is the rapid alternating of the two binary states of the light, such as *On* and *Off*. In addition, alarm lights almost always, and pilot lights sometimes, are joined with an audible device—a horn, buzzer, chimes, or even a recorded vocal message that names the process condition—to attract the operator's attention. The operator then looks for the light that shows what has happened.

Pilot lights and alarm lights in a plant are frequently color coded to convey more information than merely whether a particular piece of equipment is *On* or *Off* or whether a condition is normal or abnormal. The color code may show relative importance or priority in case of trouble, designate the process system that is involved, or provide other information.

Pilot lights to denote whether a piece of equipment is operating are often combined with the manual switches that control the equipment. For example, a rotary two-position switch or a double push-button switch can light up a *Steam On* or *Steam Off* light, according to which switch position or button is selected.

3-3-1-4 Annunciators

An annunciator is a device or a packaged group of devices that call attention to changes in process conditions that have occurred. An annunciator usually signals abnormal process conditions but may also be used to signal normal process status. It usually calls for interaction with the plant operator.

The attention-getting signals that an annunciator sends out are usually of two kinds:

1. *Visible*. An annunciator cabinet usually has a rectangular group of small boxes with windows that are translucent, white or colored. They are engraved with service descriptions, and the engravings are usually blacked in so they can be read when the windows are lighted from the back. A typical window inscription might be as follows:

   SUMP LEVEL HIGH

   LAH-215

   (LAH-215 is the instrument number for high-level alarm 215. See Section 3-2-6.)

2. *Audible*. An audible alarm is provided in the control room to supplement the visible alarm, as described in Section 3-3-1-3, "Lights." Additional audible alarms may also be placed in other plant locations.

Annunciators have an operating logic, which is a built-in program that controls how visible and audible output signals of the annunciator are turned on by process trouble and how they are turned off. This logic is known as an *operating sequence*. Very many different sequences can be purchased. ISA standard S18.1, *Annunciator Sequences and Specifications*, outlines a method for designating sequences, but it does not name any sequence as being standard. Two popular sequences are ISA Sequences A and F1A, which are described in Figure 3-10.

Annunciators are usually located in a plant control center but are also found in local areas of the plant. Their operating logic may be located in the annunciator cabinet but is often placed in a remote cabinet, especially for large systems.

A so-called *first-out annunciator* is useful where a single process failure may cause a chain of other seeming failures to occur; this results in a group of alarm windows being lighted almost at the same time. What should the operator look at first? The first-out annunciator directs him or her to the alarm that started the trouble because only that window is flashing; the succeeding alarm windows have steady lights.

For example, such a progression could happen if there is a loss of cooling water, which causes a failure to cool lubricating oil, which in turn causes a compressor to stop on high oil temperature, which then causes a loss of compressor discharge pressure. Several alarms would go on, but the real culprit is the failure of cooling water.

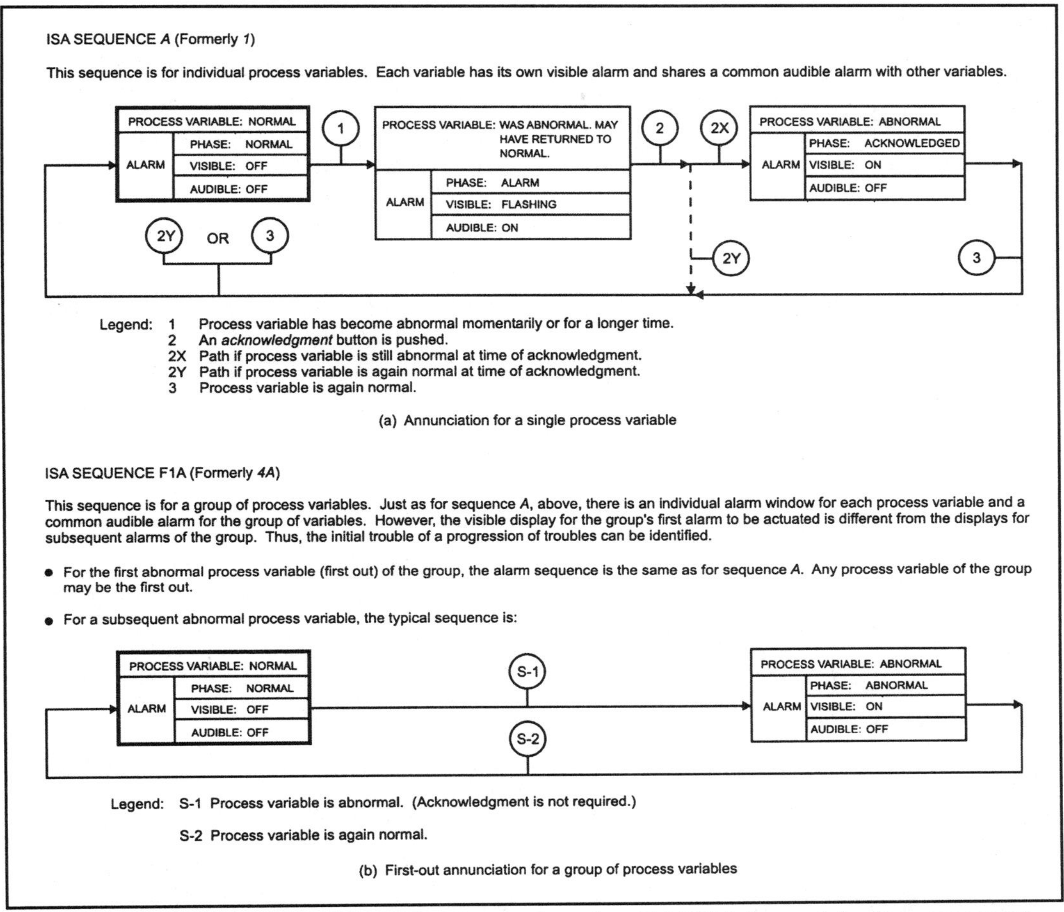

Figure 3-10. Two Popular Annunciator Sequences (based on ISA S18.1 Annunciator Sequences and Specifications.)

Annunciators may be provided with many optional features in addition to those illustrated in Figure 3-10. Among these features are the following:

1. *Ringback.* This signals the operator when the process variable for the alarm returns to normal.
2. *No Lock-In.* This causes the annunciator sequence to return to the normal state without requiring the operator to acknowledge the abnormal state, provided that the process variable has become normal.
3. *Automatic Alarm Silence.* This silences the audible alarm automatically after a fixed time even though the alarm may not have been acknowledged. The visible alarm is not affected.
4. *Auxiliary Binary Electric Output Signals.* These may be used to operate *slave alarms* or other remote devices. A slave alarm copies the alarm action of the main alarm, but usually in a simplified sequence.

Pneumatic annunciators with pneumatic operating logic are available, but they are used infrequently in the process industries.

3-3-1-5 Video Screens

The video screens can display any kind of information that can be put on paper: drawings, lists, instructions, and so on. Video may be found with computer control and distributed control and programmable logic control (see Section 6-4-2).

3-3-2 Presenting and Retaining Permanent Information

Permanent information may be presented by chart-type recorders, printers, or other recording devices. These are described in the following sections.

3-3-2-1 Chart-type Recorders

The most common type of process recorder is essentially a machine that draws graphs on a paper known as a chart. The most common type of graph shows how a measured variable, such as flow, varies with time. This type of recorder uses a clock drive to move the chart while an actuator that responds to a flow signal moves a pen across the chart. The chart provides analog information for current use by the operator, and it may be filed for future reference.

Charts in a control room are usually rectangular, often four inches or twelve inches wide and in long lengths that are typically adequate for a month's continuous use; these are termed *strip charts*. Strip charts may be used in the field, but many charts for field use are usually round, twelve inches in diameter, and are usually good for twenty-four hours of use, sometimes for seven days or a month.

Recorders may be either single-point type or multipoint type, depending on the number of measured variables that are to be recorded. A single-point instrument serves one variable, such as pressure or temperature. A multipoint instrument may serve a variety of many variables.

The single-point recorder has a single pen. The multipoint recorder may have either one or many pens. The points of a multipoint recorder may share the same chart space, and their lines may cross each other; each of the various records is identified and distinguished from its companions by a color code or printed numbers. On a twelve-inch strip chart there is usually only one pen, which switches among twenty-four input signals. (This is an example of a multiplexing device, as described in Section 3-2-5.) Some charts carry eight or more parallel pens in separate channels without overlap.

Chart recorders are available with alphanumeric readings and with auxiliary output signals for switching or controlling.

Data loggers are often used to keep track of all the important variables in the plant. They record operating data, perform calculations, and check for alarm conditions. They record routine operating data by switching internally from point to point; however, they give recording priority to abnormal conditions and then switch to high-speed printing.

The record they provide is an alphanumeric tabulation, which contains data such as the date and time, point identification, service, and value of the process variable. The printing is normally black but may be switched to another color in order to highlight abnormal conditions.

Event recorders perform some of the same functions as data loggers. They may normally be kept in a standby state, ready to start up immediately upon the occurrence of an abnormal condition, and they record the abnormalities. Their printing speed is very high, with times reported down to a hundredth of a second. This makes it possible to see what conditions initiated a sequence of alarms (see the above discussion of first-out annunciators in Section 3-3-1-4 and Figure 3-10).

3-3-2-2 PRINTERS

Printers are furnished as integral parts of certain recording systems and are also furnished routinely as so-called *peripheral devices*, auxiliaries for computers and distributed control systems. They may also be used to draw graphs of data stored in a computer or distributed control system.

3-3-2-3 OTHER RECORDING DEVICES

A variety of other means are available for storing operating information. These include the following:

1. *Magnetic tape.* The information may be analog or digital. It cannot be read directly.
2. *Video tape.* This is similar to magnetic tape but can also store pictures and sounds.
3. *Video copier.* This can photographically copy whatever a video screen displays.
4. *Computer disks.* These include 5 and 1/4-inch and 3 and 1/2-inch floppy diskettes as well as large-memory hard disks.

3-4 HOW PRESSURE SENSORS SENSE

3-4-1 UNITS OF PRESSURE AND FORCE

Pressure and *force* are very important factors for industrial processes and the world we live in. It may be helpful to review the difference between these two terms.

If we place a 10-inch cube of metal weighing 260 pounds on a table top, the cube imposes a downward force of 260 pounds on the table. The contact area of the cube with the top is 10 inches by 10 inches, or 100 square inches. The pressure, however, is the force divided by the contact area, 260 pounds divided by 100 square inches, or 26 pounds per square inch (psi). This downward pressure is not on the entire top but only on the 10-inch-by-10-inch portion of the top. If the top were 4 feet square—2304 square inches—and were covered solidly with 10-inch-high pieces of metal, the downward force on the table would be 60,000 pounds, or 30 tons. The table would have to be very special not to collapse under that load, but the pressure on the tabletop would still be only 26 psi (60,000 pounds divided by 2304 square inches).

If a person blows air through his or her lips, he or she creates a small, low-pressure wind with a force that can deflect a sheet of paper and not much more. An atmospheric wind with the same low pressure can move sailing ships because the pressure is exerted over the large area of sail, thereby creating a large force.

Figure 3-11 shows two vertical pipes on supports, one pipe having an inside diameter of 2.0 inches, the other an inside diameter of 8.0 inches. The bottom of each pipe is closed, the top open. Both pipes are filled with cold water to a level of 27.68 inches. For each pipe, the downward force on its support is the same as its weight of water (ignoring the weight of the pipe). The force is 3.14 pounds for the 2-inch pipe and 50.28 pounds, sixteen times as much for the 8-inch pipe. But the pressures at the bottom of both pipes are identical at one psi; that is because the pressure depends on the height of water, not the volume. The height of liquid is known as *head*.

The head of atmospheric air on the Earth extends to the stratosphere and beyond. The weight of the atmosphere and its force on the entire surface of the Earth are colossal but on one square inch of soil at sea level the force is approximately 14.7 pounds; therefore, the pressure is 14.7 psi.

In the United States, we use various units to express pressure in engineering work. Most used is pounds per square inch, symbolized as *psi*. To measure very low gas pressures, we use *inches of water*, symbolized as either *in. WC* or simply *WC*, both meaning *water column*. Measured inches of water is the height or head of water whose bottom pressure equals the pressure of the process variable sensed by the instrument (see Figure 3-12). One inch of cold water creates a force of 0.0361 psi.

Pump pressures are frequently expressed as *feet of head*, shown as *ft head*. This refers to a theoretical column of the pumped liquid that the pump pressure supports. The height of the theoretical column, the pump head, changes as the density of the liquid increases or decreases. *Density* is the ratio of the mass of a substance to its volume. *Mass* is a basic measure of the quantity of matter.

Thus, oil floats on water while stones sink because of their different densities. The density effect is often significant when head is the sensed variable used for making measurements of flow, level, or other variables.

Barometric pressure is frequently expressed in *inches of mercury* (in. Hg) or in *millimeters of mercury* (mm Hg). Here we refer to a column of liquid mercury. Measured at sea level, barometric pressure is normally 29.92 in. Hg, equal to 760.0 mm Hg and 14.70 psi. This is the so-called *normal barometer*, whose frequent changes affect all of us and also gage-pressure instruments (see Section 3-4-2).

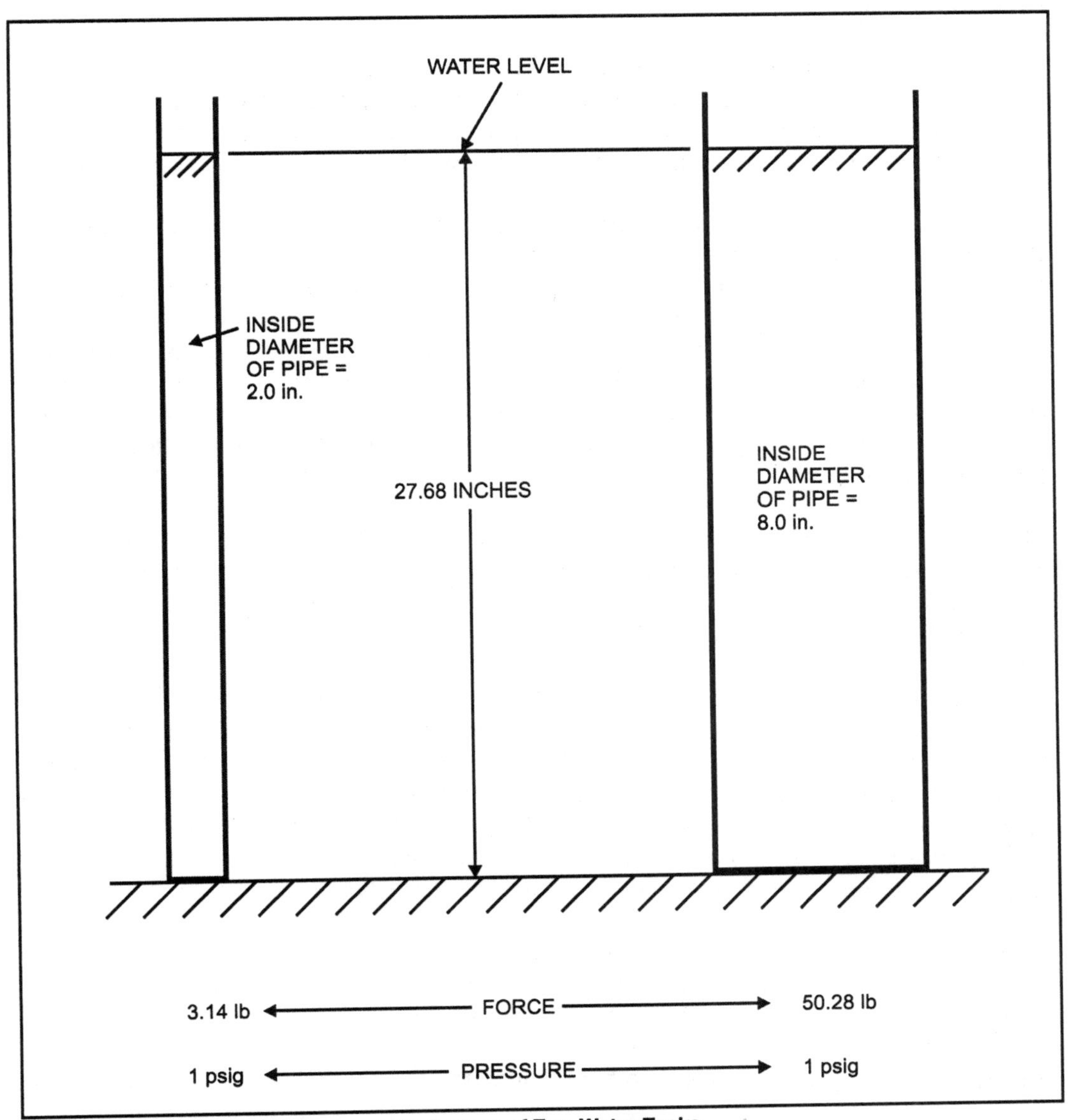

Figure 3-11. Force and Pressure at the Bottom of Two Water Tanks

For force, the United States uses the engineering unit *pound*, meaning *pound avoirdupois* (lb), which has sixteen ounces to the *pound*. *Ton* is also used, meaning the *short ton*, which equals 2000 pounds.

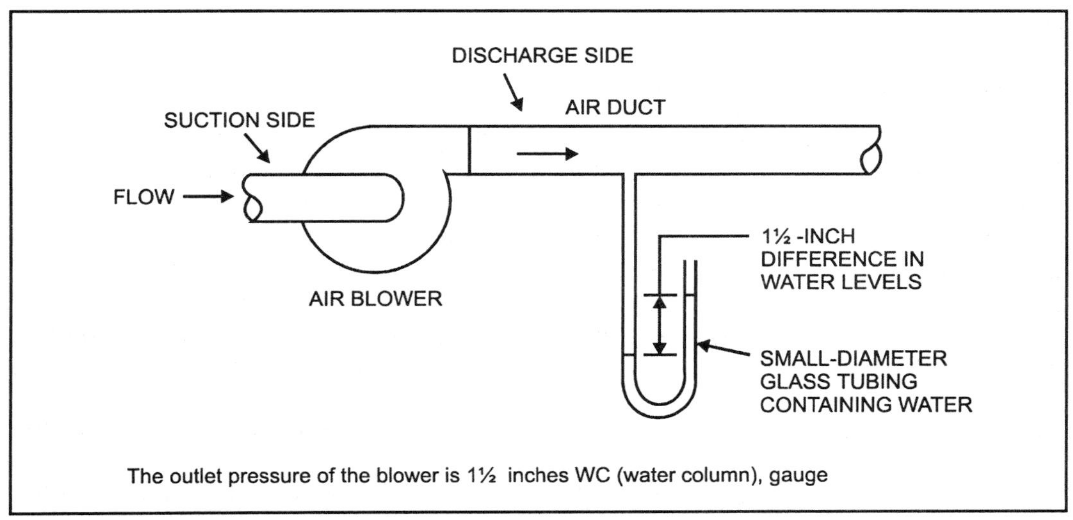

Figure 3-12. Measuring Low Pressure

In the United States, units of force and of weight are the same except that an object weighing 10 pounds at sea level exerts a downward force of 10 pounds but, when raised to a higher elevation, the force is less than 10 pounds. This is because the weight of an object is really the downward pull on the object caused by gravity; the farther the object gets from the center of the Earth, the less is the downward force and the corresponding weight. What remains unchanged about the object is its mass. However, we are seldom concerned about mass when we deal with process measurement and control. We usually deal with weight. Nevertheless, there are cases where gravity effects may be significant (see Section 9-2-6).

A helter-skelter system of measuring units used to exist throughout the world and still exists in the English system that we continue to use in the United States. In 1791, the French Academy of Sciences established the metric system, a new, sensible system of units using decimal relationships. The metric system spread throughout the world. In 1960, thirty-six countries, including the United States, modified the metric system and established the *International System of Units*, which is internationally symbolized as *SI*. The SI system is the world standard. U. S. practice does not yet make full use of SI or of its predecessor, the metric system, but there is a slow but definite trend in the United States toward the SI system.

The SI unit of pressure is the *pascal* (Pa). There are variants, such as *kilopascal* (kPa, a thousand pascals), *millipascal* (mPa, a thousandth of a pascal), and so on. A pascal equals one newton per square meter; a kilopascal equals 0.1450 psi. The *newton* (N) is the unit of force, independent of gravity, and equals 0.2248 pounds of force. The *meter* (m) (sometimes spelled *metre*) is the unit of length and equals 39.37 inches.

The metric unit of force is the *gram*, which is now replaced by the newton. The metric unit of pressure is *kilogram per square centimeter*, which is replaced by the pascal.

3-4-2 KINDS OF PRESSURE

Figure 3-12 shows pressure being measured for process air whose pressure is higher than atmospheric pressure. The higher pressure pushes the water down on one side of the measuring instrument while the water is pushed up against atmospheric pressure on the other side; the difference in water levels is 1 1/2 inches. The pressure measurement is made using atmospheric pressure as a reference. Whenever the reference pressure is atmospheric, the measured pressure is called *gage pressure*. (*Gage* is sometimes spelled *gauge*.) Therefore, the pressure on the discharge side of the blower is 1 1/2 inches of water gage.

The gage pressure may be either positive (that is, above atmospheric as in Figure 3-12) or negative (below atmospheric). A negative gage pressure is a vacuum that, if it existed in the process, would suck the water of the instrument higher on the left-hand side, lower on the right-hand side.

If all the air were removed from the tubing on the atmospheric side—the right-hand side—of the instrument and if the open end were sealed tightly, then the reference pressure would be absolute zero instead of atmospheric. This would give us an *absolute-pressure* measurement, which is always positive.

The instrument in Figure 3-12 is connected to the discharge side of the air blower. If we now connect the open end of the instrument to the suction side of the blower, the instrument will sense the difference in pressure, the *differential pressure* or *pressure differential*. In print, these terms may be symbolized by P or ΔP. The instrument would show how much the blower increases the pressure of the air in going from blower suction to discharge. The blower is powered by an electric motor or some other type of drive and imparts energy to the air passing through the blower. This energy is a process input.

The kind of instrument we have illustrated to measure blower pressure is a *U-tube manometer*. Its various uses are shown in Figure 3-13. Note the basic similarity of cases (a), (b), (c), and (d) in Figure 3-13; in all four cases, the right-hand side of each manometer is connected to a process pressure. The essential difference among the cases is the nature of the reference pressure on the left-hand side: the left-hand side is connected to the atmosphere in cases (a) and (b), to a total vacuum in case (c), and to a second process point in case (d). In cases (a) and (b), an increase in atmospheric pressure pushes the liquid down on the left side of the manometer so the liquid on the right side moves up, thereby reducing the reading in (a) and increasing it in (b); this reflects the fact that a given physical pressure of the process gives a gage-pressure reading that depends on the atmospheric pressure, which is changeable. The relationship between absolute and gage pressures is shown in Figure 3-13(e).

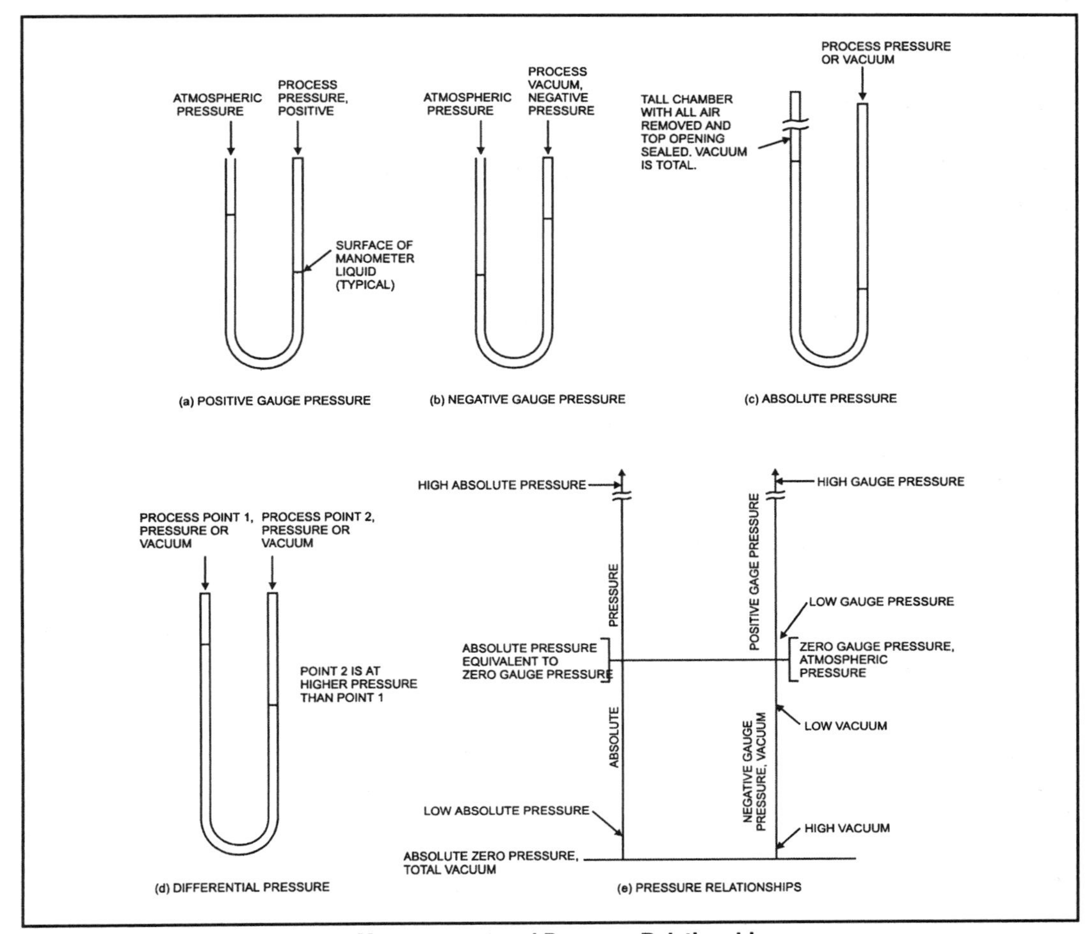

Figure 3-13. Types of Pressure Measurement and Pressure Relationships

The word *pressure* is sometimes ambiguous. It is used in a generic sense to denote any of the following: positive gage pressure; negative gage pressure, which is a vacuum; absolute pressure; or differential pressure. It is also used specifically to denote a positive gage pressure as opposed to a vacuum. Thus, we have a pressure regulator that prevents a tank from both bursting from pressure and collapsing from vacuum. It is very common for pressures to be stated merely in psi; this can sometimes lead to uncertainty and error. Generally, the best and safest practice is to use the pressure letter symbols listed here, especially if there is a chance of misunderstanding as to the type of pressure measurement:

Gage pressure	psig
Absolute pressure	psia
Differential pressure	psid

Vacuums and pressures stated in terms of head of liquid are usually understood to be gage pressures, but if there may be misunderstanding a clarifying word or symbol may be helpful.

3-4-3 Types of Pressure Sensors

The basic design of several common types of pressure sensors is described in this section. Some of these instruments may indicate the measured pressure. They are often used for other functions, including transmission, switching, controlling, and recording.

3-4-3-1 Manometer

The principles of manometers for measuring pressure have been covered in Section 3-4-2. The glass U-tube manometer is used for pressures and vacuums that are close to atmospheric pressure. Metal manometers are used for higher pressures. The manometer is important because it is one of the types of highly accurate measuring standards used for calibrating plant instruments (see Section 3-1-4).

3-4-3-2 Bourdon

The most widely used pressure instrument in a plant is usually the pressure gage, which is used for local measurements. The most common sensing element among all types of industrial pressure instruments is the Bourdon tube, which is illustrated in Figure 3-14.

The Bourdon tube was patented by Eugene Bourdon of France in 1840. The tube is made of metal, is curved, has an oval cross section, and acts as a spring. The inside of the tube is at process pressure, and the tube is inside a case that is generally at atmospheric pressure. Thus, the tube senses the difference between the pressures that are inside and outside the tube; therefore, the difference is the gage pressure. As the process pressure increases, the cross section tends to become less oval, more circular, and the tube tends to straighten out. This causes the far end of the tube to move. The movement, through links and gearing, causes a pointer to rotate around a circular graduated dial. There are variations on the basic design.

A round-dial pressure indicator used to measure tire pressure is apt to have this type of construction.

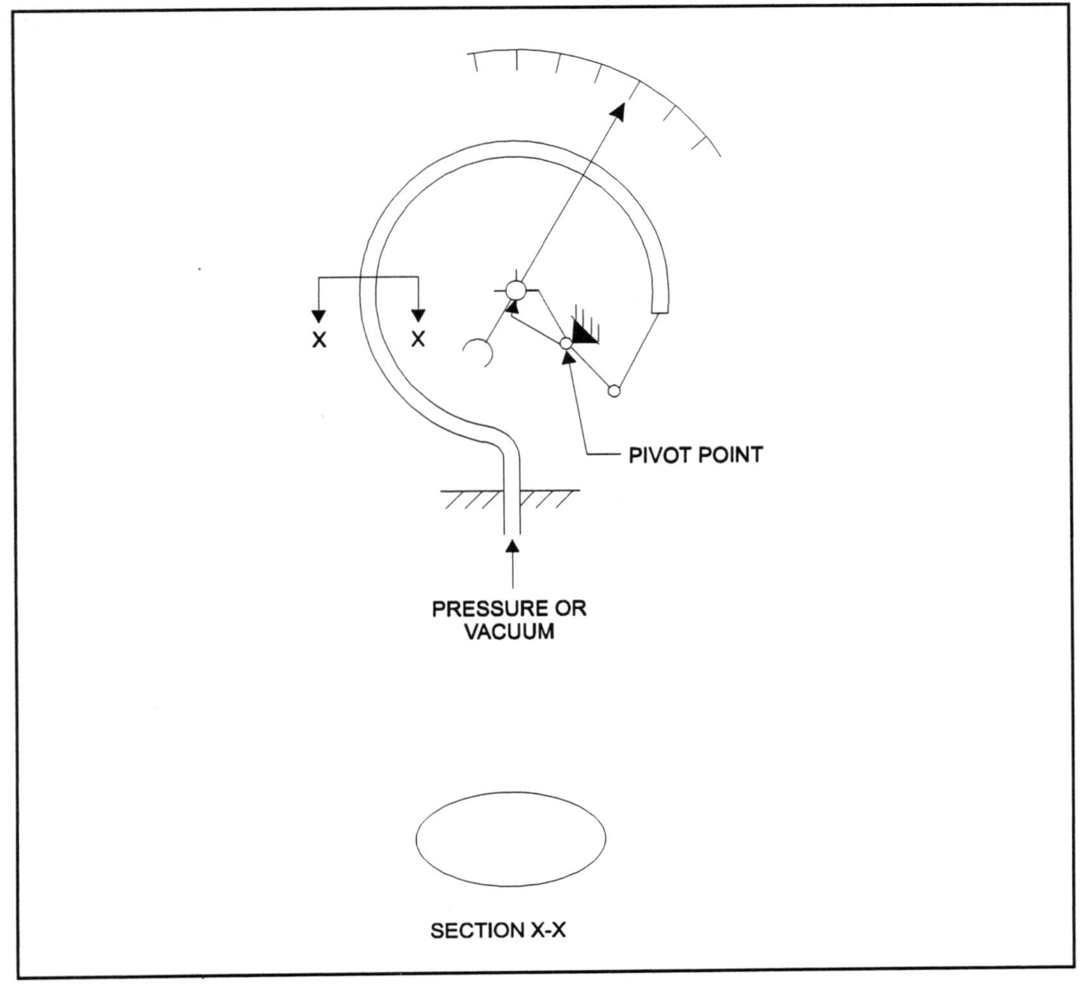

Figure 3-14. Basic Design of Bourdon Sensor

3-4-3-3 Bellows

Another common type of sensor uses a metal bellows or capsule that performs somewhat as the Bourdon sensor does. Pressure contracts the bellows; release of pressure permits the bellows to relax. Movement of the bellows is communicated to a pointer, which permits us to read the pressure.

Figure 3-15 shows how a bellows-type pressure instrument is modified to measure absolute pressure and differential pressure. Other types of pressure sensors can be similarly modified to measure any type of pressure, assuming that the sensitivity and structural strength of the sensor fit the application.

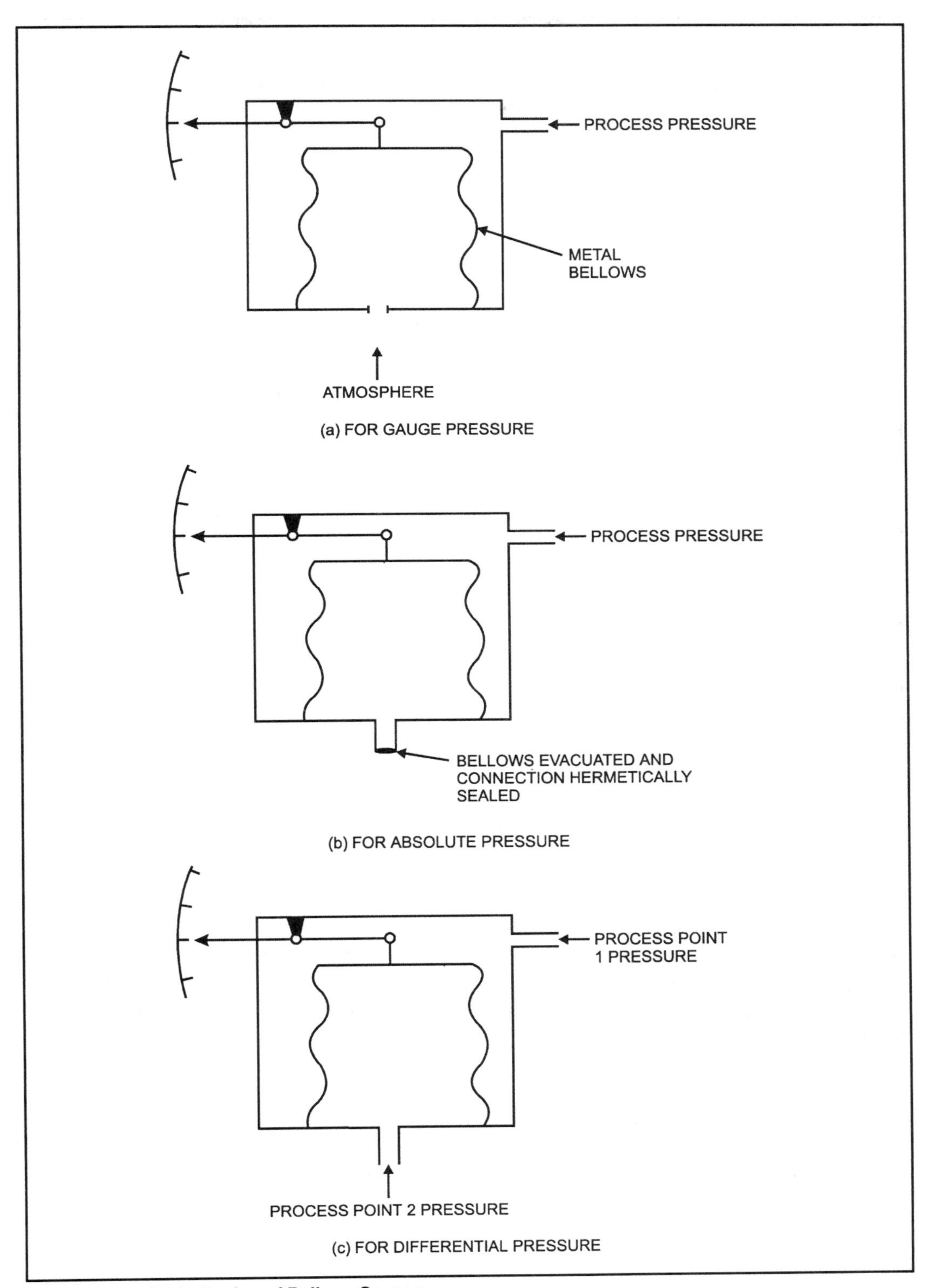

Figure 3-15. Basic Design of Bellows Sensor

3-4-3-4 Diaphragm

The force from the process pressure acting against a flexible diaphragm is opposed by a spring. Increasing pressure results in the movement of a pointer. This is illustrated in Figure 3-16.

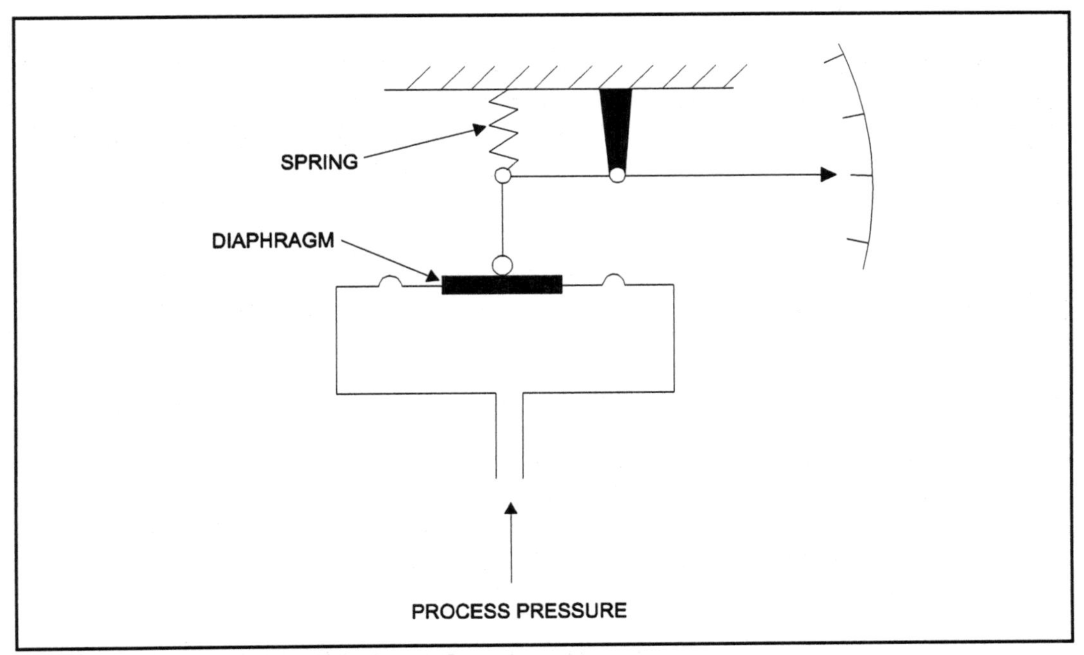

Figure 3-16. Basic Design of Diaphragm Sensor

3-4-3-5 Strain Gage

When a solid object is subjected to pressure, it deforms and is said to undergo *strain*. A strain-gage sensor, or simply *strain gage*, is an electronic device with an electrical capacitance or resistance that changes as the sensor undergoes strain. When the sensor feels a process pressure, the sensor's capacitance or resistance can be measured and correlated with pressure.

3-5 How Flow Sensors Sense

3-5-1 Units of Flow

In the United States at present, the English system of units is in general use for plant measurement of process flows. The units depend on the physical state of the flowing substance: gas, vapor, liquid, or solid (particles). For flow rates, the units are commonly as shown in Table 3-4. Note that some of the measuring units are based on volume, some on weight. However, all flow measurements may be based on weight. The unit *gallon* in the table refers to the U.S. gallon, not the British imperial or the Canadian gallon.

The word *flow* is ambiguous. It usually means *flow rate,* but, used loosely, it sometimes means *volume,* as in "stop the pump after a flow of 1000 gallons."

Table 3-4. Flow Measuring Units

Fluid State	Units
Gas (see Appendix B)	Cubic feet per hour (cfh), cubic feet per minute (cfm)
Steam and other vapors (see Appendix B)	Pounds per hour (pph)
Liquid	Gallons per minute (gpm), barrels per day (bpd) in the petroleum producing and refining industries (1 barrel = 42 gallons), cubic feet per second (cfs) in the water and sewage industries
Solid (powders or lumps)	Pounds per hour (pph), tons per hour (tph)
Slurry (liquid-and-solid mixtures)	Pounds per hour (pph), tons per hour (tph)

The SI system uses the following units for flow rates:

- *Volume-based.* Cubic meters per second (m^3/s). (1 cubic meter = 264.2 gallons. The volume unit *liter*—sometimes spelled *litre*—is recognized by SI, but its use is discouraged; *cubic decimeter,* which is equal to liter for practical purposes, is preferred. One cubic decimeter equals 0.2642 gallons.)
- *Mass-based.* Kilograms per second (kg/s). (1 kilogram = 2.205 pounds.)

The volume-based units for gas flow, such as cfm, have a significant refinement, resulting in units of *scfm* or *acfm,* meaning *standard cubic feet per minute* or *actual cubic feet per minute,* respectively. The actual volume of a given gas flow depends on the existing temperature and pressure of the gas. The actual volume is converted to a standard volume by calculating what the volume would be if the gas were at a standard temperature and a standard pressure. Thus, a meter that measures the flow of fuel gas to the home senses acfm, but the monthly bill we receive is based on that measurement corrected to state the scfm. We pay for the correct amount of gas we consume, neither more nor less.

The so-called standard conditions for gas measurement, which are referred to as *base conditions,* must be defined for each contract or project because there are no "standard" conditions. Base pressures that are used include 14.4, 14.65, 14.70, and 14.9 psia. Base temperatures include 32, 60, 68, and 70°F. The mass or weight of a given type of gas in a "standard" volume depends on which base conditions are used. The differences between the base conditions and the distinction between acfm and scfm can be important for a plant that buys or sells any gas, purchases equipment such as a gas compressor, or checks on plant performance.

An analogous situation regarding standard conditions exists for the volume-based measurement of liquid flow. There are *standard gallons* and *flowing gallons*. However, temperature is the major influence on liquid volume; the effect of pressure is usually unimportant. In the United States, the most commonly used base temperature for liquids is 60°F. Gasoline stations measure the volume of gasoline delivered to a car at the flowing temperature, but the measurement is automatically corrected to a standard temperature, and we pay accordingly.

Mass-based flow measurements are not sensitive to temperature or pressure and are generally not subject to misunderstanding. The same is true, practically speaking, for weight-based measurements. However, significant changes due to changes of gravity may lead to erroneous measurements unless we compensate for the gravity effects.

3-5-2 Types of Flow Sensors

There are a bewildering number of sensor types to measure the fluid flow rate and volume. Many sensors can be grouped into several different types, as described in the following sections.

3-5-2-1 Differential Pressure

This type of flow sensor is the one most commonly used in industrial work. The most common of the type is the *orifice plate* or *thin-plate orifice*, which is usually a circular thin metal plate having a concentric hole with a sharp inlet edge. The plate is set into the pipeline.

Wherever there is an orifice plate—or any other restriction or obstacle to straight flow—the fluid pressure downstream of the plate is lower than the upstream pressure. This difference in pressure created by the plate is called *differential pressure*, which has a value of zero at zero flow and increases as the flow increases. An instrument, such as an indicator or transmitter, that can sense the differential pressure can then be a flow meter. The orifice plate is a *primary element*, and the indicator (or transmitter) is used as a *secondary element*; together they measure the flow rate.

Figure 3-17 shows the orifice plate installed in the pipeline and the resulting streamlines, which are the paths followed by the fluid stream. The streamlines bend with the pressure changes in the pipe, and the figure shows how the secondary instrument is connected to sense differential pressure, thereby measuring the flow.

The figure also shows a partial recovery of pressure, which leaves a permanent decrease of pressure from the fluid constriction, just as a kitchen faucet that is only partially open restricts the flow and causes the water to come out as a weak stream; this decrease is called *pressure loss*.

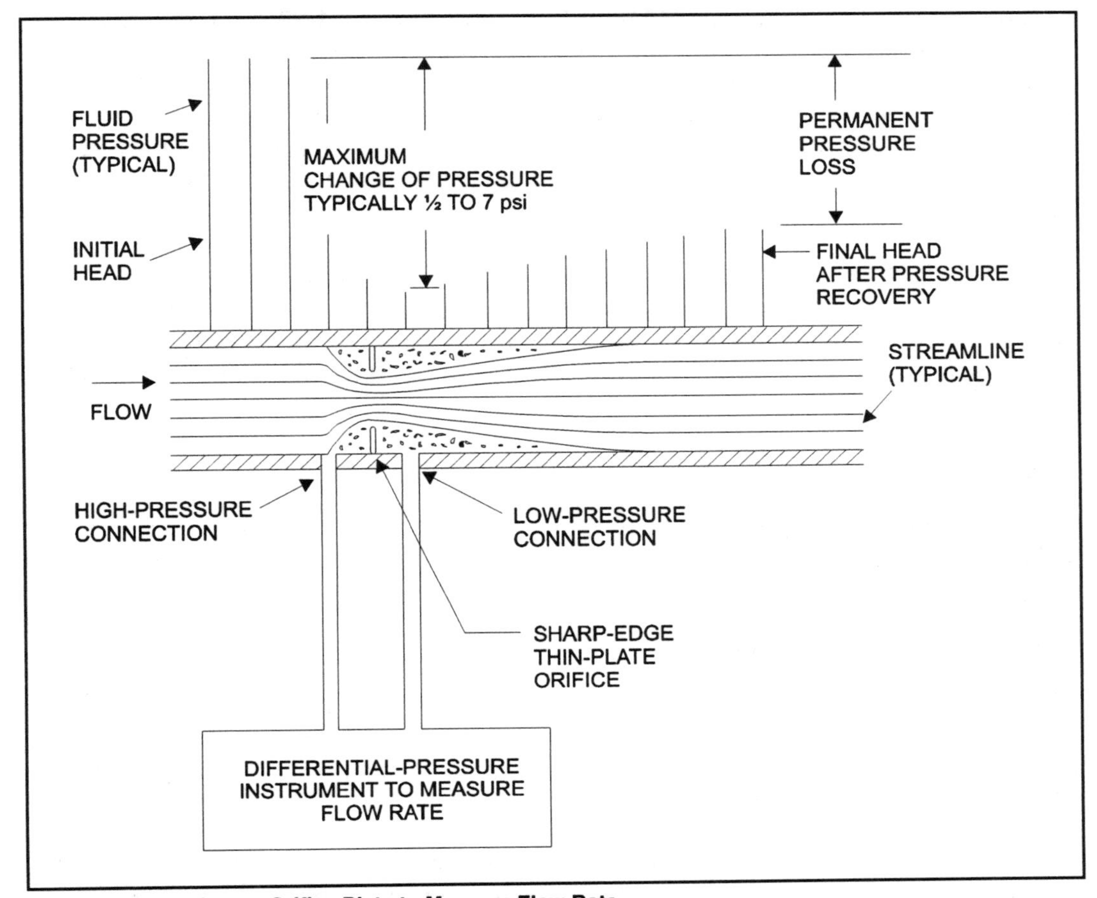

Figure 3-17. Using an Orifice Plate to Measure Flow Rate

Mathematical equations are used to size the hole accurately according to the fluid, the maximum flow rate, the flow conditions, and the exact locations of the differential-pressure-sensing connections on the pipe. There are several standard pairs of locations for the upstream and downstream connections, which are called *taps*. The most common arrangement uses *flange taps*, for which the center of the connections is one inch before and one inch after the respective faces of the plate. Other arrangements are known as *radius taps, vena contracta taps, pipe taps,* and *corner taps.*

The plate depicted in Figure 3-17 may have variations: greater thickness, an off-center location for the hole, a noncircular hole, a curved instead of sharp-edged inlet for the hole, different metals for the plate—all for specialized purposes. Sometimes a specially shaped tube—known as a Venturi tube, (after the 18th- and 19th-century physicist G. B. Venturi)—or one of its variants is used instead of an orifice plate. The tube works on the same principle as for the plate.

For a differential-pressure flow element to measure with maximum accuracy, the flow must have a *normal velocity profile* or a *fully developed velocity profile,* which is a symmetrical flow pattern of the fluid all around the pipe. If a run of pipe is straight without internal disturbance and of sufficient length, then the velocity profile is normal. Otherwise the profile is distorted, and the flow measurement is less accurate. Recommended minimum lengths of straight pipe to provide a normal profile are established by the American Society of Mechanical Engineers (ASME) and the American Gas Association (AGA).

In order to meet the accuracy requirement for flow measurement, a pipe run may have to be lengthened. This would increase the cost and may be physically difficult to do. An expedient sometimes used to shorten the straight-run requirement is to install straightening vanes, which are a short length of a stack of small pipes, or the equivalent, inserted inside the main pipe.

Another widely used type of differential-pressure flow element for general process use is the averaging-type Pitot tube, which is a modern modification of the traditional Pitot tube, invented by the Frenchman Henri Pitot in the 19th century. Essentially, the Pitot tube now uses two small tubes that enter the side of the main pipe and extend across it. One tube has holes facing upstream, the other has one hole facing downstream; so the tubes have a differential pressure that depends on the flow rate. A secondary instrument connects to the tubes and is the flow meter.

For many kinds of instruments, the output varies proportionally with the input. A percentage change in one is matched by the same percentage change in the other. This is generally true for pressure indicators, for which doubling or tripling the input doubles or triples the reading. This relationship is called *linear* or *uniform*.

The same is true for many other instruments, including many flow instruments. However, it is not true for differential-pressure flow instruments, for which the output varies as the square of the flow. For example, if an orifice plate gives an output differential pressure of 20 inches WC for a 300-gpm flow, then doubling the flow to 600 gpm quadruples the output to 80 inches WC. Tripling the flow causes the output to be nine times as large. This is an example of a *nonlinear* or *nonuniform* relationship. Unless this nonlinearity is altered by a special instrument, it requires a square root nonuniform measuring scale. (For more on measurement nonlinearity, see Figure 3-9(d) and Table 4-2, lines 4 and 11.)

3-5-2-2 ROTAMETER

A rotameter, also known as a *variable-area* meter, has a vertical tapered tube, small end down, with fluid flow upward. Inside the tube is a weight that rests at the bottom of the tube but that is forced higher and remains suspended as the flow increases. The weight is therefore called a *float*. It is sometimes called a *rotor* even though it does not necessarily rotate.

The tube may be of glass with graduations to permit direct reading of the flow rate, but the tube is often made of metal for strength. The float may have a vertical stem attached on top to permit it to be connected to secondary functions, such as transmission (see Figure 3-18). The rotameter scale varies proportionally with flow, and is therefore linear.

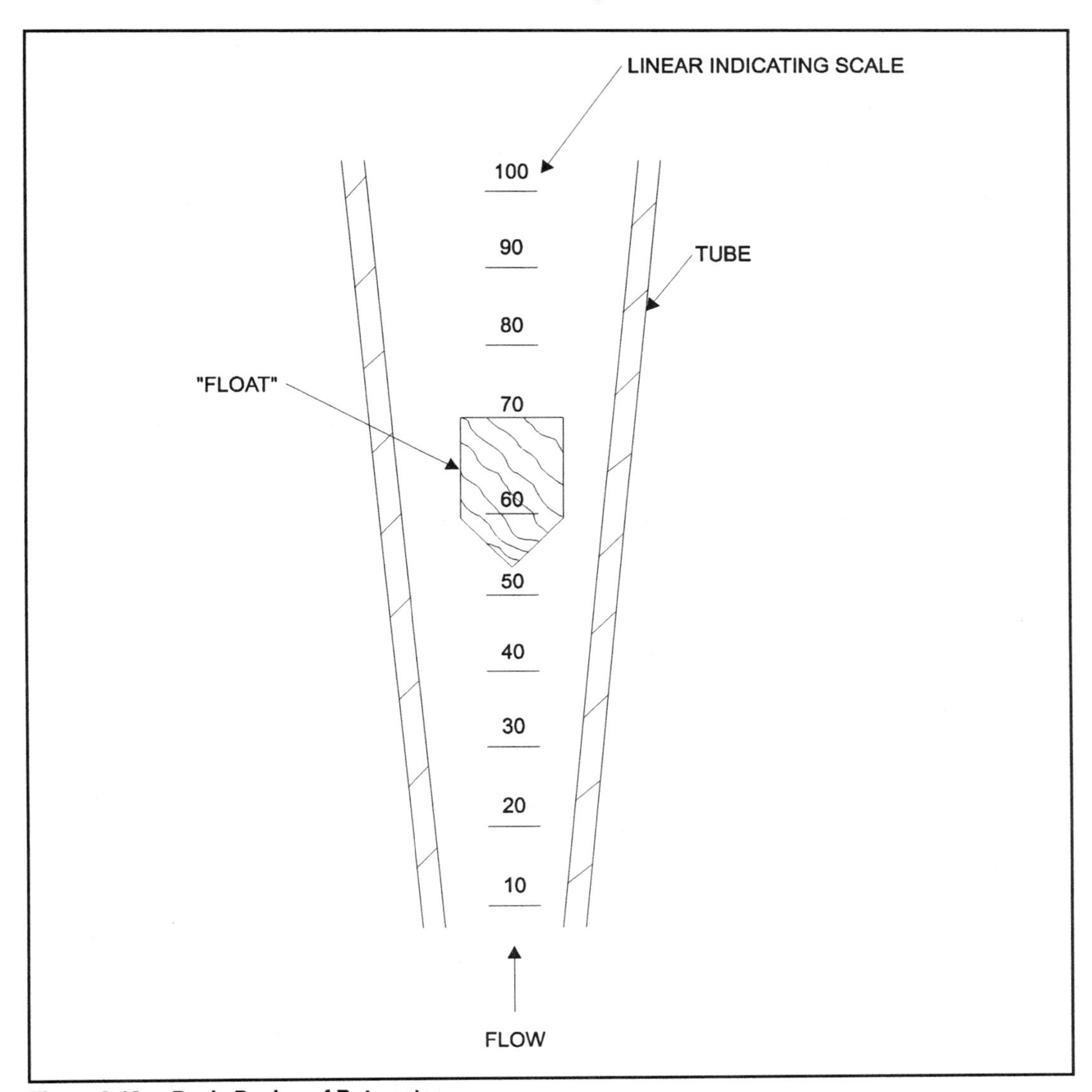

Figure 3-18. Basic Design of Rotameter

3-5-2-3 Other

Other types of flow sensors are used in process plants, and each type has its own set of pros and cons for different applications. Included among the other sensors are the following:

1. Positive displacement, like the water or gas meter used in homes.
2. Magnetic.

3. Turbine.
4. Vortex.
5. Thermal.
6. Ultrasonic.
7. Weir and flume, to measure flow in open channels.
8. Weight (for flow rate of solids on a conveyor).
9. Coriolis.
10. Ultrasonic time-of-transit.
11. Doppler.
12. Radar.

The positive-displacement and turbine sensors rotate and can be supplied with a transmitting element that sends out a pulse with every rotation of the sensor. This output signal is digital and can be used directly or be converted to other forms.

3-6 How Level Sensors Sense

3-6-1 Units of Level

Level measurements are usually expressed as a distance or as a percentage of level span but are sometimes stated as a volume or weight. For height, the units may be inches or feet, whichever is more convenient. The SI unit for height is generally the meter.

But what shall be the datum, the starting point, for the measurement? Consider the tank of Figure 3-19, which is a composite of different situations. The level-measuring instrument has a range whose lower and upper limits may be chosen differently according to the situation and differently by different engineers. Real possibilities are the following:

- For the lower limit of measurement

  Bottom of normal band

  Bottom of overall operating band

  Centerline of side outlet connection

  Inside bottom of side outlet connection

  Lower tangent line of tank

  Extreme bottom of inside of tank

- For the upper limit of measurement

    Extreme top of inside of tank

    Upper tangent line of tank

    Centerline of overflow connection

    Inside bottom of overflow connection

    Centerline of side inlet connection

    Top of overall operating band

    Top of normal band

The combination of limits that is chosen depends on their significance to the process and to the plant operators. The limits should be clearly defined to avoid uncertainty or misunderstanding as to what a given level reading really means.

Whatever the measuring range may be, the upper sensing point must be at the same elevation as or higher than the upper limit of the range, and the entire lower sensing line must be at the same elevation as or lower than the lower limit of the range. Otherwise, some of the range will not be usable.

Let us choose a measuring range from the bottom of the side outlet to the bottom of the overflow connection and say the span is eight feet. A level indicator could be calibrated either 0 to 100% or 0 to 8 feet, using the lower limit of the range as the reference level or datum; or -50 to +50% or -4 to +4 feet, using the midpoint of the range as the datum; or 353 to 361 feet, using the elevation of the plant as a datum. All these ranges are equivalent, but the most meaningful one for the application should be selected.

A level instrument may be calibrated in terms of volume, such as gallons or barrels, or in terms of weight, generally as pounds. These calibrations are simple for a cylindrical vertical tank because liquid volume and weight vary uniformly with level. If the same tank were placed horizontally, the relationship of volume and weight to level would be very nonlinear because of the curvature of the tank. Both cylindrical and horizontal styles are used.

In the SI system, the volume units for level may be cubic meters, or cubic decimeters, and the weight units would be newtons (comparable to the U. S. term, pounds of force).

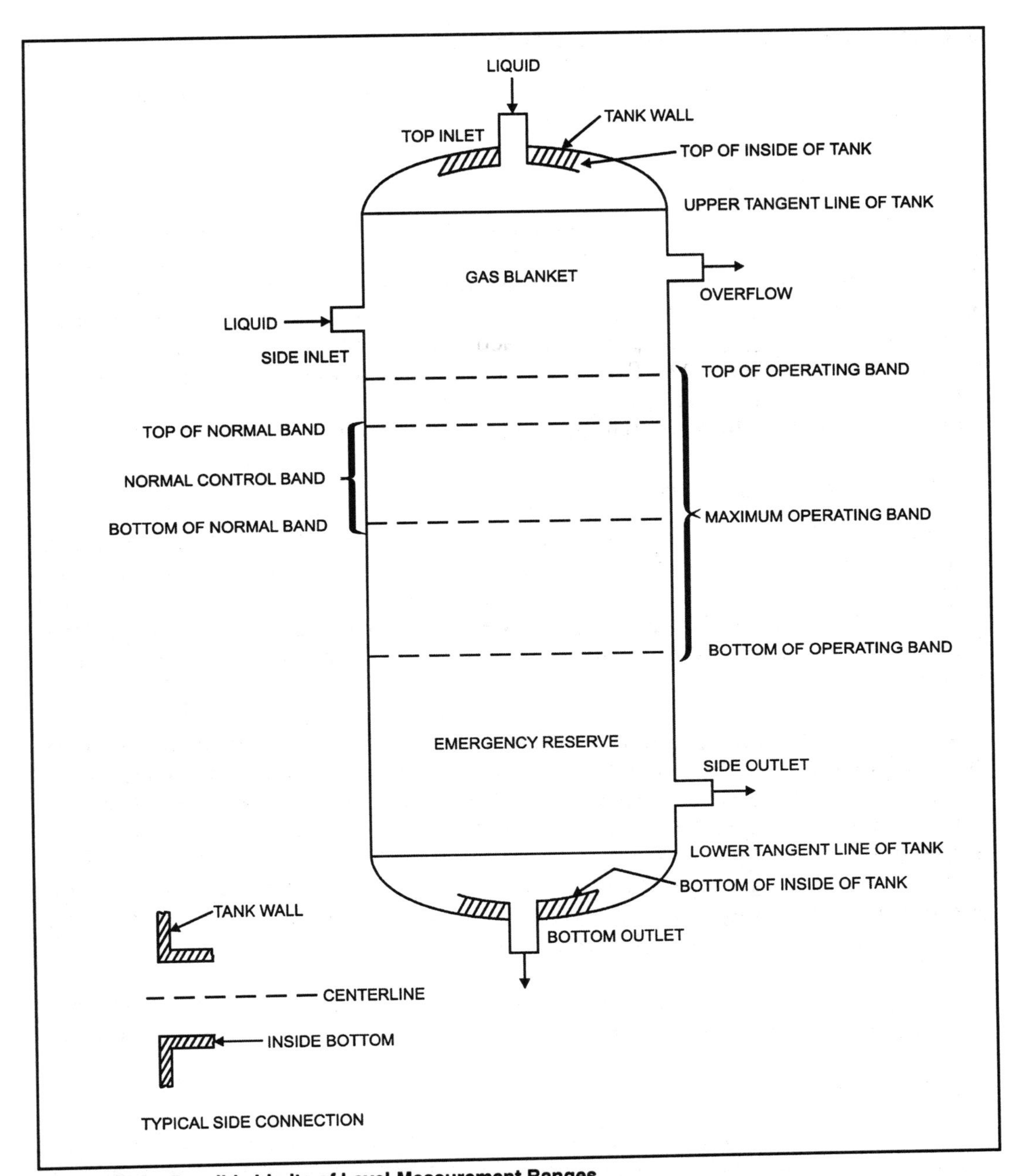

Figure 3-19. Possible Limits of Level-Measurement Ranges

3-6-2 Types of Level Sensors

There are many types of level-sensing instruments, among which are the following.

3-6-2-1 Float

One of the major types of level sensors uses a float on the surface of the liquid in a tank, which rises and falls with the level. The device used to refill a toilet tank in the home has a float sensor. Figure 3-20(a) shows examples in which a float operates (a) an electric switch when the level is above or below some fixed value and (b) a level indicator. The plant float action can provide level indication, transmission, and so on. The float may be mounted inside the tank, but, more often, it is placed inside a metal chamber that is piped to the tank and fills with liquid to nominally the same height as exists in the tank.

3-6-2-2 Displacement

Displacement-type level sensors use a vertical cylinder called a *displacer*, which is sometimes erroneously called a float. While a float stays on the liquid surface, a displacer always tends to sink to the bottom of the tank because it is heavier than the liquid it displaces. Figure 3-20(b) illustrates how the weight of the displacer is counterbalanced by a spring. The displacer, like the float, may be mounted inside or outside the tank.

When the level is at the bottom of the range, the spring bears the entire weight of the displacer, and the indicator shows zero level. A transmitter output signal would be at the bottom of its range. As the level rises, the displacer becomes partially submerged, and a buoyant force reduces the apparent weight of the displacer by the weight of liquid that is displaced by the displacer. When the level reaches the top of the range, the displacer is fully submerged, the apparent weight of the displacer is minimum, the indicator shows 100% level, and the transmitter output signal is at the top of its range.

The buoyant force follows Archimedes' Principle, which the Greek philosopher first stated in the third century B.C.

3-6-2-3 Head

The head of liquid in a tank is equivalent to a pressure. By measuring the pressure, we can know the liquid level. The pressure is measured using the pressure instruments described in Section 3-4 and shown in Figure 3-20(c). (The convention used for identifying such instruments is to label them as level instruments, according to their use, not as pressure instruments, according to their construction.)

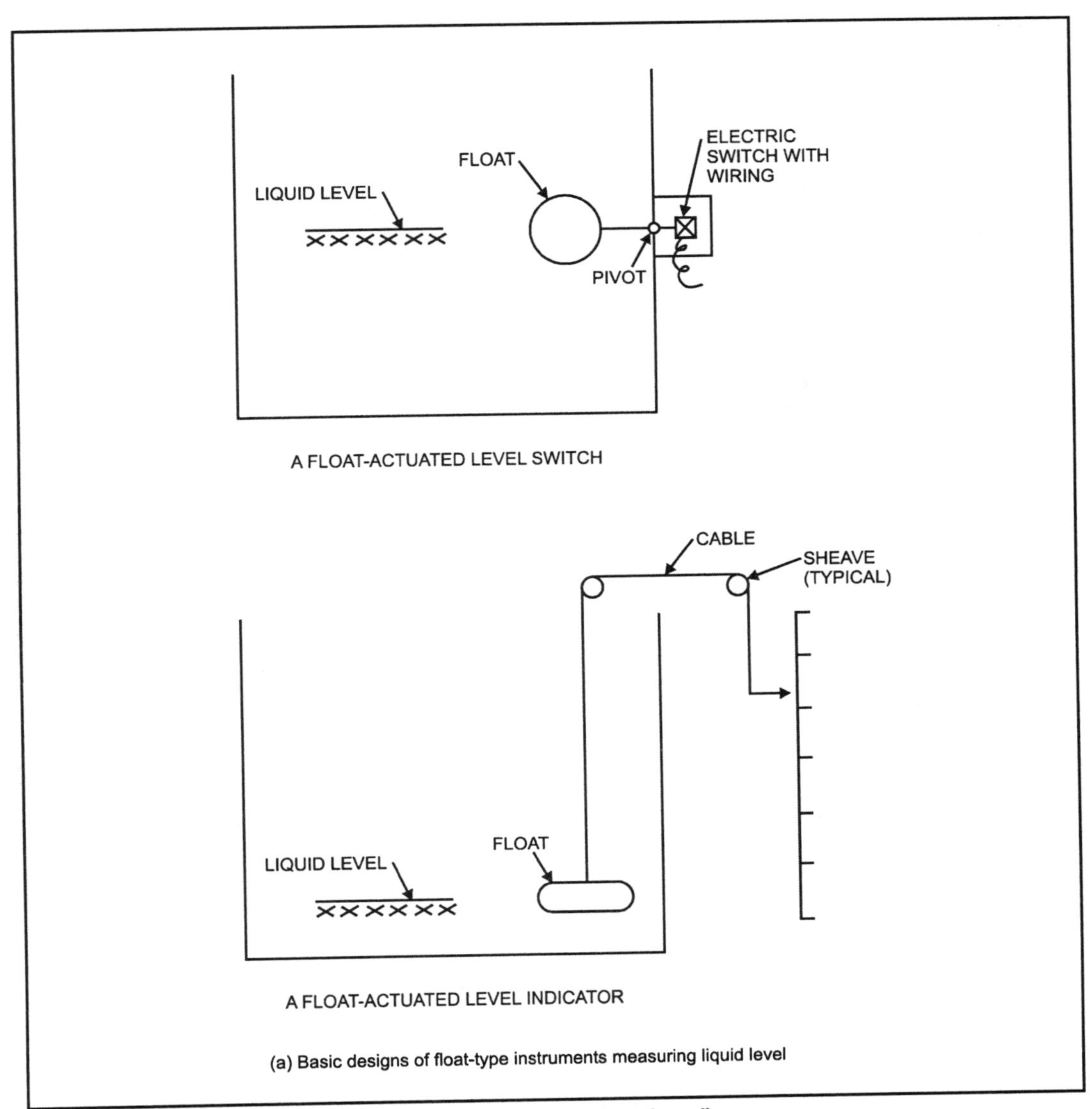

(a) Basic designs of float-type instruments measuring liquid level

Figure 3-20. Basic Methods of Measuring Liquid Level (continued)

Figure 3-20(a) shows an open tank, which is at atmospheric pressure. The level instrument to measure the liquid head can be an ordinary pressure indicator or other instrument that measures gage pressure. However, this indicator cannot be used for the closed and pressurized tank of the figure because the reading could be seriously false. The indicator would sense not only the head of liquid but also the pressure above the liquid inside the tank and would indicate the sum of the pressures.

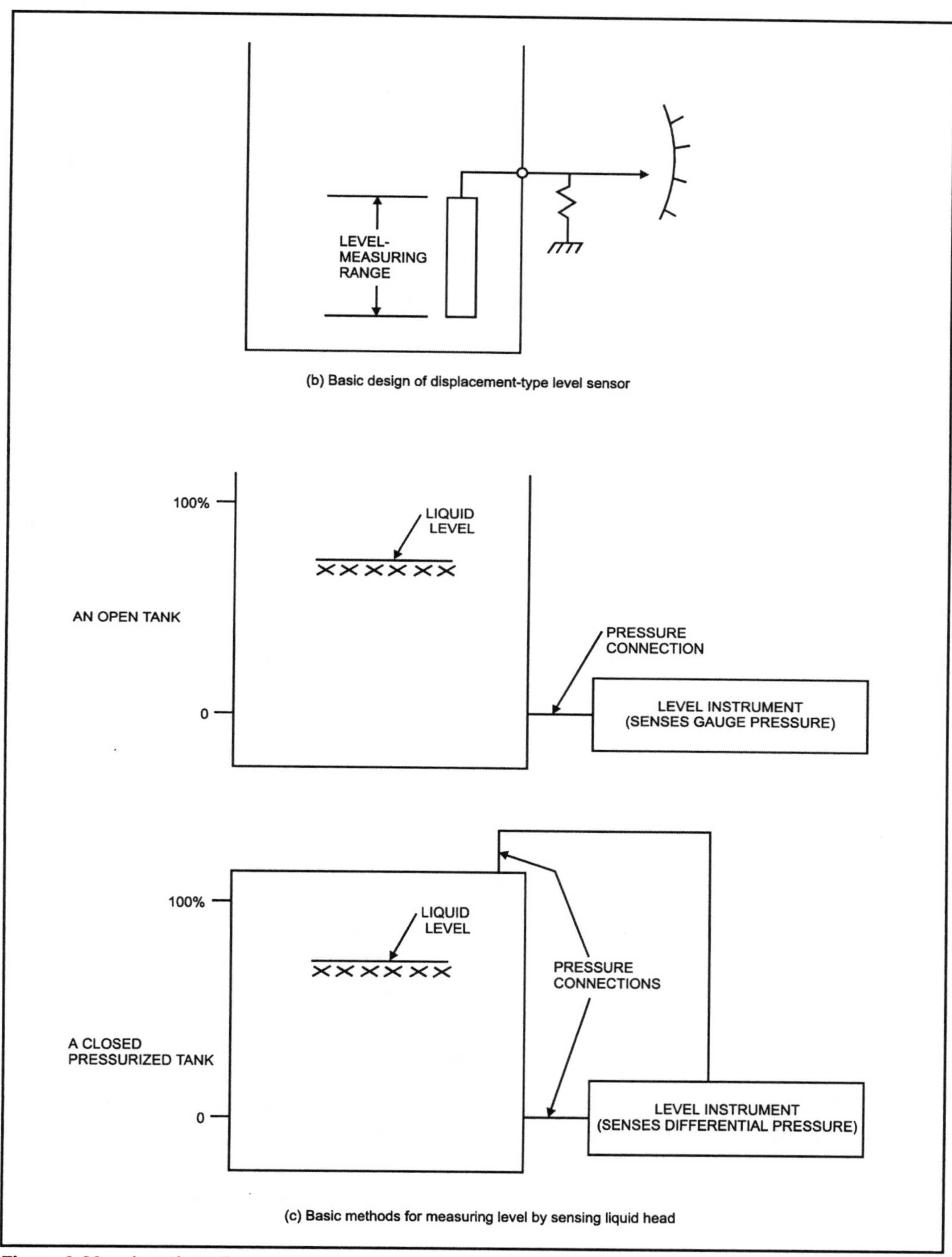

(b) Basic design of displacement-type level sensor

(c) Basic methods for measuring level by sensing liquid head

Figure 3-20. (continued)

To measure only the head, we need to use a differential-pressure instrument. We connect one side of this instrument to sense the liquid head plus tank pressure, the other side to sense only the tank pressure. The instrument subtracts the tank pressure from the sum of head plus tank pressure, and the result is pure liquid head. If the tank level is the same in the two tanks of Figure 3-20(c), both their instruments will give the same reading.

3-6-2-4 Glass

A basic level instrument that is widely used, very often in conjunction with other types of level instrument, is a *level glass, level gage,* or *gage glass*. This is a vertical length consisting of either glass tubing or a metal chamber with one or two glass windows. The tubing or chamber is piped to the tank and shows nominally the same liquid level as exists in the tank.

3-6-2-5 Other

Other types of level sensors include the following:

1. Capacitance probe. A *probe* is a slender rod that is inserted into the process fluid.
2. Conductance probe.
3. Radiation, using a nuclear source.
4. Ultrasonic.
5. Optical.
6. Thermal.
7. Rotating paddle, used to measure solids level in a bin.
8. Weight. A weight scale senses the weight of the tank plus its contents but can generally be adjusted to subtract the tank weight. The result is the weight of the contents.

3-7 How Temperature Sensors Sense

3-7-1 Units of Temperature

In the United States, the temperature in process systems is usually expressed in degrees Fahrenheit, which is symbolized by °F and is named after Gabriel D. Fahrenheit, a German physicist of the 17th and 18th centuries. The Fahrenheit scale is based on the freezing point of water, 32°F, and the boiling point of water, 212°F at normal atmospheric pressure. The lowest possible temperature, absolute zero, is at -459.69°F. An absolute temperature scale, in *degrees Rankine* (°R), has degrees Rankine being equal to degrees Fahrenheit plus 459.69. Thus, a room temperature of 72°F equals 531.69°R. The unit degrees Rankine is named for the 19th-century Scottish physicist William J. M. Rankine.

In the SI system, the usual temperature scale is in *degrees Celsius*, after Anders Celsius, the Swedish inventor of the centigrade scale in the 18th century. What used to be *degrees centigrade* should now be called degrees Celsius, symbolized by °C. This scale defines the freezing point of water as 0°C and the boiling point as 100°C. The corresponding absolute temperature scale in *kelvins* begins at absolute zero, or -273.15°C. Thus, 20°C equals 293.15 K. The kelvin is named for the 18th- and 19th-century Irish-British physicist William T. Kelvin.

One degree Celsius equals 1.8°F. Thus, the number of degrees Celsius equals the difference: degrees F minus 32, divided by 1.8. Thus, -40°F equals -40°C.

3-7-2 Types of Temperature Sensors

Most temperature sensors that are inserted into a process fluid are provided with a protective enclosure (see Section 10-1 on temperature wells in Chapter 10).

3-7-2-1 Bimetallic

The most common type of sensor for reading a process temperature locally is a bimetallic thermometer (Figure 3-21 [a]). This sensor uses two thin strips of different metals bonded together over their length. When the bonded metals are heated, they expand unequally and the strips bend toward the side that expands less. The bending causes movement of the end of the bimetal, which results in a pointer moving to indicate temperature.

The same bimetallic principle is used to actuate electric switches and is found in electric irons and thermostats in the home.

3-7-2-2 Thermocouple

A thermocouple consists of a pair of wires joined at one end, with each wire being of a different metal. When the joined end is heated, a small electrical voltage can be measured at the open end of the pair.

Voltage is an electrical force, usually stated in the unit of *volt* (V) or *millivolt* (mV, one-thousandth of a volt) for thermocouples. Voltage is the force that pushes electrons (subatomic electrical particles) through an electrical circuit. The flow of particles is a *current*, usually measured in the unit *ampere* (A) or *milliampere* (mA). The material property that impedes the flow of current is resistance, usually measured in the unit ohm (Ω). The volt is named for the Italian Alessandro Volta, the ampere for Andre M. Ampere of France, and the ohm for the German Georg Simon Ohm; all were 18th- and 19th-century physicists.

The joined end of the thermocouple is called the *measuring junction;* the open end is the *reference junction*. The voltage increases as the heated end gets hotter, so measuring the voltage of a thermocouple submerged in a process fluid is equivalent to measuring the temperature of the fluid. The basic measuring system, using a millivoltmeter to measure the voltage, is shown in Figure 3-21(b).

The thermocouple response is the Seebeck effect, named after Thomas J. Seebeck, a German physicist who discovered the effect in the 19th century.

The measuring junction and reference junction are sometimes referred to as *hot* and *cold junctions*, respectively, but this usage is not recommended. The words *hot* and *cold* can be misleading because thermocouples can be used to measure very low process temperatures, in which case the "hot" junction is colder than the "cold" junction.

Standards have been established for the type and composition of the metals and alloys that are used for thermocouples. A commonly used type of thermocouple is *iron-constantan* (ISA Type J), which has one wire of pure iron and another of constantan. Constantan is a specific alloy of nickel and copper.

3-7-2-3 Resistance

Another important class of temperature sensors is the *resistance type*. This type depends on the fact that the electrical resistance of a wire increases as the wire temperature increases, as first described by the German-British inventor Sir William Siemens in the 19th century. If the electrical resistance of a given type of wire submerged in a process fluid is measured, that is equivalent to measuring the temperature of the fluid.

The sensors may be made of various metals, in which case they are known as *resistance temperature detectors* (RTDs) or *resistance thermometers*. Pure copper, nickel, and platinum are widely used for RTDs, and a metal alloy is also used. The sensors may also be made of metal oxides and are then known as *thermistors* (THERMal plus resISTOR).

3-7-2-4 Filled Thermal System

Filled-system temperature sensors have a small metal bulb that is connected to a pressure sensor by means of small-bore tubing. The whole system is filled with a fluid and is then hermetically sealed. If the bulb is submerged in a hot or cold process fluid, the fluid in the bulb gets hot or cold, and the pressure in the filled system rises or falls. The pressure sensor responds to the change in pressure by, for example, moving a pointer to indicate the new temperature (see Figure 3-21[c]).

The filled system may be filled 100% with a liquid, such as mercury, or a gas, both of which expand as they are heated. Or it may be partially filled with a volatile liquid whose vapor pressure increases as the liquid is heated; the remainder of the space is filled with the vapor. *Vapor pressure* is a measure of the tendency of the liquid to evaporate. In each case, the filled-system pressure increases with increasing process temperature. The thermal characteristic of the fill fluid is known, so the indicator has temperature graduations marked accordingly.

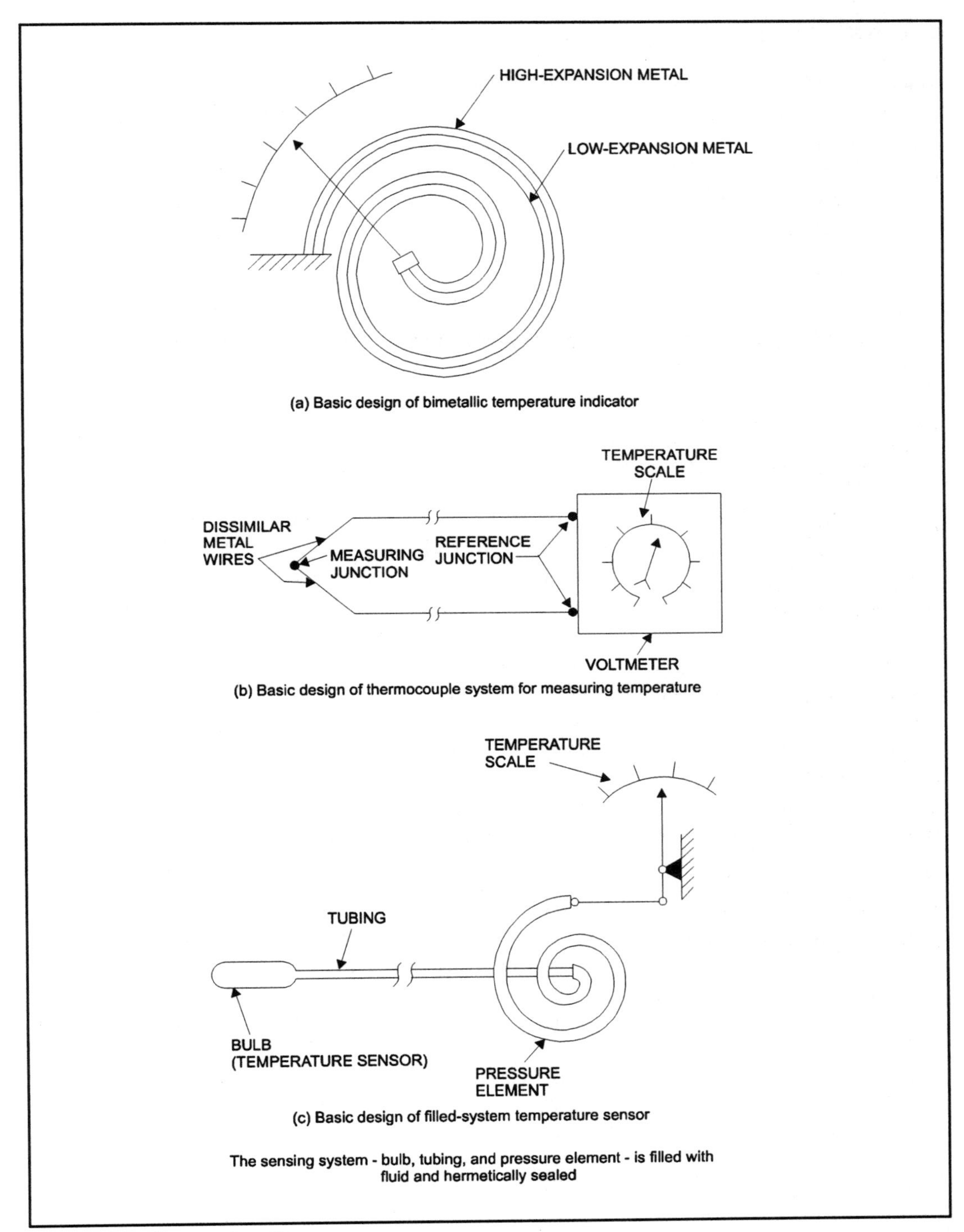

(a) Basic design of bimetallic temperature indicator

(b) Basic design of thermocouple system for measuring temperature

(c) Basic design of filled-system temperature sensor

The sensing system - bulb, tubing, and pressure element - is filled with fluid and hermetically sealed

Figure 3-21. Temperature Sensors

3-7-2-5 Other

In addition to the four types of temperature sensors just described, there are the following:

1. *Pyrometer.* (This is for very high temperatures like those of molten metal or the inside of a furnace.)
2. *Color-change paint.*
3. *Fusible shapes.*

3-8 How Analysis Sensors Sense

3-8-1 Units of Analysis

Analysis sensors, or simply *analyzers*, determine the composition of a *mixture* or a *solution* of different substances or the concentration of a particular substance.

A mixture is an intermingling of two or more different substances that are not bound to each other and that retain their individual characteristics. The substances merely share a common space. Examples are a mixture of water and petroleum oil, water and sand, and fly ash in the smoke of a fire. Mixtures can be treated to separate their components, for example, by filtering the sand out of the water.

A solution is an intermingling and merging of two or more different substances so that their individual characteristics disappear and the different substances are not distinguishable; the solution is uniform throughout. For example, let us add a small quantity of table salt to a glass of pure water. The white salt dissolves and disappears from view. The liquid looks unchanged but it has become a solution, and its properties have changed. Its freezing point is lowered, its boiling point is raised, it conducts electricity better, its density is raised, and its viscosity and still other properties are changed. Continue to add salt. A point is reached where no more salt can dissolve, and we say that the solution is now *saturated*. Add a tiny bit more salt; it will not dissolve but will fall to the bottom. We now have (a) a salt-and-water solution and (b) a mixture of solid salt with the solution. We can filter out the crystals from the mixture. If we then heat the solution, the water boils off and leaves behind all the salt that had been dissolved.

Analyses are generally stated as a *concentration* in one of the following ways:

- Percent by weight.
- Percent by volume.
- Parts per million (ppm).
- Parts per billion (ppb).

An important exception in stating analyses is the measure of acidity, which depends on the *activity* of hydrogen ions in a liquid. (Water is made up of hydrogen ions and hydroxyl ions.) Activity is related to concentration and is usually stated as *pH*, which stands for the *power of the hydrogen ion*.

Pure water at specified conditions has a pH of 7.0 and acids have a pH of less than 7.0, while alkalies—the inverse of acids—have a pH of more than 7.0. The pH of orange juice is approximately 3, of milk 6, and of magnesia solution 10.

3-8-2 Types of Analysis Sensors

Analysis measurements are inferential, that is, they depend on properties of the particular substance we are analyzing (see Section 3-1-1). A measurement of these properties tells us the concentration of the substance. For example, the concentration of salt in a water solution can be determined by measuring either the density of the solution, its ability to conduct electricity, optical characteristics, or other properties. In many cases, a particular analysis can be performed by more than one kind of analyzer.

The analyzers mentioned in the next five sections are used for continuous analysis of a process and, aside from servicing requirements, operate without human attention. There are also laboratory-type analyzers and techniques that are for use by a control chemist in a laboratory.

3-8-2-1 Infrared

Infrared rays have frequencies lower than those of the red end of visible light and higher than those of microwaves. With few exceptions, when infrared radiation has a frequency that oscillates over a range, some of the radiated energy is absorbed by chemical molecules and increases the motion of the molecules. For different chemicals, the molecules absorb different combinations of frequencies of the radiation. The pattern of absorbed frequencies identifies the molecules in the sample. The amount of absorption of the rays at specific frequencies is a measure of the concentration of the molecules.

Infrared analyzers are used for gases and liquids.

3-8-2-2 Thermal Conductivity

Thermal conductivity is the ability of a substance to conduct heat. Different substances conduct heat at different rates. In practice, what is determined is not the actual thermal conductivity of the sample in question but rather its conductivity compared to that of air at specified conditions. At 212°F, air has a conductivity of 1.000, and the conductivity of ammonia, for example, is 1.084.

Thermal-conductivity analyzers are used for gases and vapors.

3-8-2-3 Acidity-Alkalinity Sensing

As explained in Section 3-8-1, the degree of acidity or alkalinity of a liquid is stated by the pH of the liquid. The pH is measured by a probe that uses a special kind of glass that is sensitive to hydrogen ions. This glass is wetted on the outside by the process liquid and on the inside by a solution of constant pH. This probe, which has a wire immersed in the internal solution, is known as the *measuring electrode.* A second probe, known as a *reference electrode,* has an internal wire that is immersed in a special solution that maintains a stable electrical potential. The wires from the two probes are connected to a circuit that measures electrical voltage, which represents pH.

3-8-2-4 Chromatography

A chromatograph passes a multicomponent sample, mixed with an inert carrier gas, through a tube. The sample may be liquid, vapor, or gas, but the chromatograph converts the liquid to the vapor form. The tube is filled with absorbent materials that delay the passage of the various sample components for different time periods. A sensor detects each component as it emerges from the tube. The chromatograph can present the analysis in graphical form or digitally and can transmit an output signal.

3-8-2-5 Other

Other types of analyzers include the following:

1. Electrical-conductivity meter.
2. Mass spectrometer.
3. Refractometer.
4. Analyzers based on oxidation-reduction potential, turbidity, ultraviolet radiation, and titration.

3-8-3 Sampling Systems

An analysis system cannot be better than the process sample it analyzes. Erroneous chemical analyses caused by improper sampling may be damaging monetarily or physically, or both. This is illustrated by the following analogy.

The validity of the results of a market survey or a political poll based on a sample of the population depends on how well the sample represents the entire buying population or voting population of interest. A bad sample leads inevitably to a bad judgment. Before the 1936 presidential election, the then flourishing magazine *Literary Digest* conducted a very large nationwide poll and publicly predicted that Alf Landon would defeat Franklin D. Roosevelt for the presidency. Unfortunately for the magazine, they understated Roosevelt's vote by 19.3%, and Roosevelt was reelected in a landslide. The Gallup and Roper polls correctly predicted the winner. The magazine had used mailing lists of telephone subscribers and automobile owners, who formed a sample that was greatly *unrepresentative* of the general voting population, especially in 1932 and during the Great Depression. The magazine was so discredited that it soon went out of business.

In general, a process sampling system must perform some, but not necessarily all, of the following operations:

1. *Obtain a representative sample*. This may require the following three steps:
   a. Select a proper location for the sample point—in some cases, more than one sample point—in order to provide a representative sample.
   b. Design a sample probe for the specific application. In most cases, the probe is merely a pipe used to withdraw a process fluid sample to be sent to an analyzer. In some cases, the probe has a sensor that performs the analysis inside a process pipe or vessel; no sample is withdrawn. In either case, the probe should be placed where it will be in contact with a representative sample.
   c. Design a tubing or piping system to transport the sample to the analyzer. The possibility that the sample composition may change in transit should be taken into account. Also, the sample time lag should not be excessive for the particular application. Figure 3-22 illustrates different transportation methods: Detail (a) of Figure 3-22 is the basic method; (b), (c), and (d) are methods for decreasing the sample lag. In considering whether to conserve or dispose of the spent sample and the bypassed fluid, the economics of conserving and the environmental effects of not conserving need to be taken into account.
2. *Reduce the sample temperature* to a value that the sample system requires.
3. *Reduce the sample pressure* to a value that the analyzer and auxiliary devices require.
4. *Treat the sample* by one of the methods illustrated in Figure 3-22. Choose the one that is most appropriate.
   a. Filter out or trap debris and other undesirable solids.
   b. Adsorb or trap undesirable gases or liquids.
   c. Evaporate or condense the sample to put it into the physical form—liquid, vapor, or gas—required by the analyzer, and adjust the temperature as required by the analyzer.
   d. Adjust the sample flow rate to keep the sample lag to an acceptably small value; this may require a large flow. At the same time, the withdrawal of perhaps costly sample material and thermal energy from the process should be kept to an acceptable minimum. In some cases, the removal of an additional one-tenth of a gallon per minute of process sample beyond what is required may be surprisingly costly over a year's time.

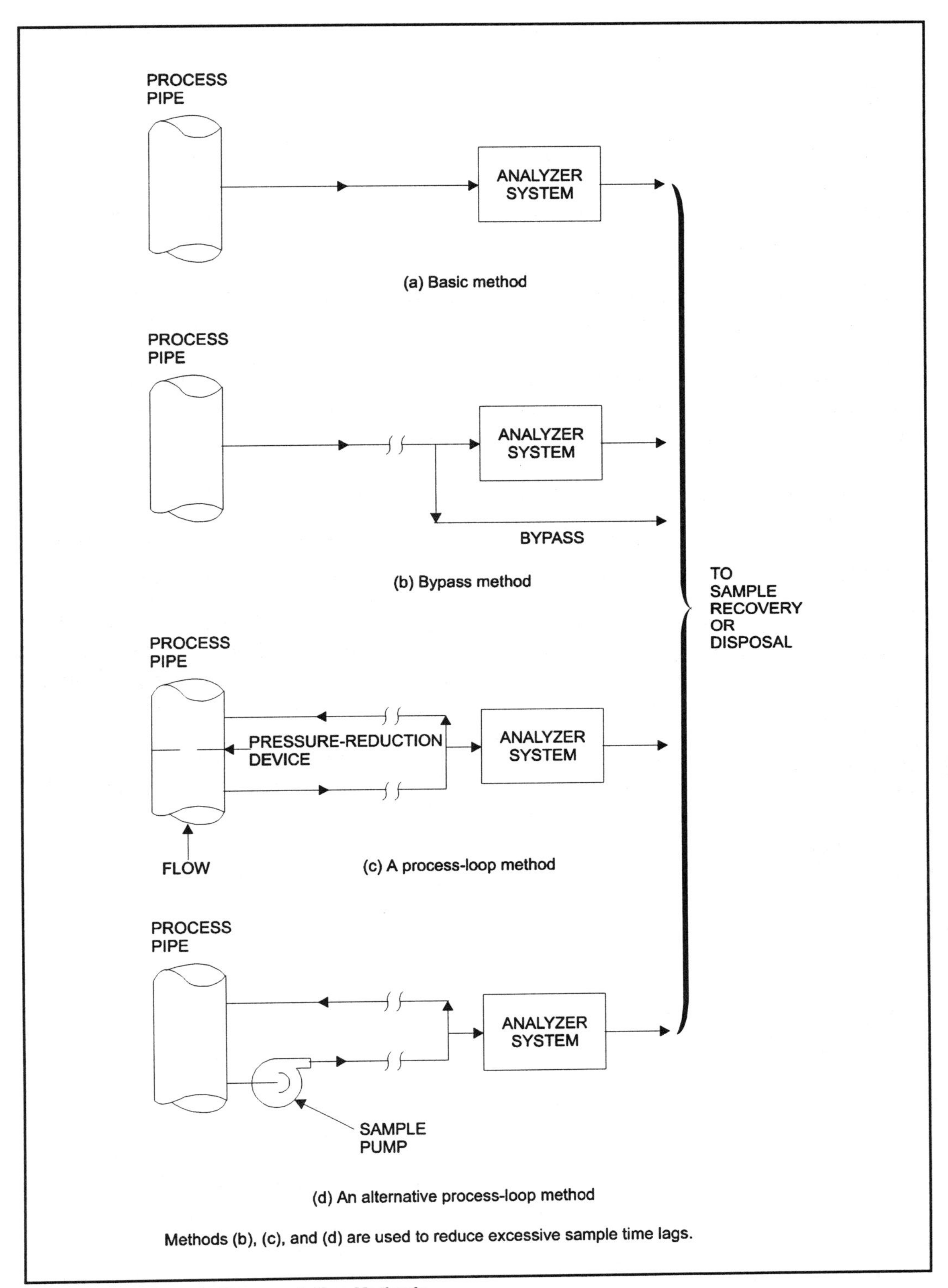

Figure 3-22. Sample-Transportation Methods

5. *Maintain the sample in the proper condition* for analysis, and deliver it to the analyzer.
6. *Recover or dispose of the spent sample* and the bypass fluid.

The whole sampling system is depicted in basic form in Figure 3-23.

3-9 Sensors for Miscellaneous Process Variables

The previous sections described representative ways to measure the most frequently encountered process variables: pressure, flow, level, temperature, and analysis. However, process plants have many other process variables that need to be measured and controlled for proper operation. These include electrical conductivity, power, time, events, position, radioactivity, vibration, viscosity, and others.

3-10 How Measurements Are Combined

Sometimes we have to obtain information that we cannot measure directly but that we can calculate from other information. Suppose we need a system to measure the rate of heat input to a furnace that burns fuel gas. If we can continuously measure or know the heating value per cubic foot of the gas as it flows and also measure the flow rate in cubic feet per hour, then the heat input rate equals the heating value per cubic foot times the number of cubic feet per hour. The result is heat input per hour. This is accomplished as shown in Figure 3-24.

Similarly, the value of other process variables can be calculated automatically and used for automatic control, the operator's information, cost accounting, and other requirements. ISA standard S5.1 includes a standard set of symbols for a variety of mathematical functions.

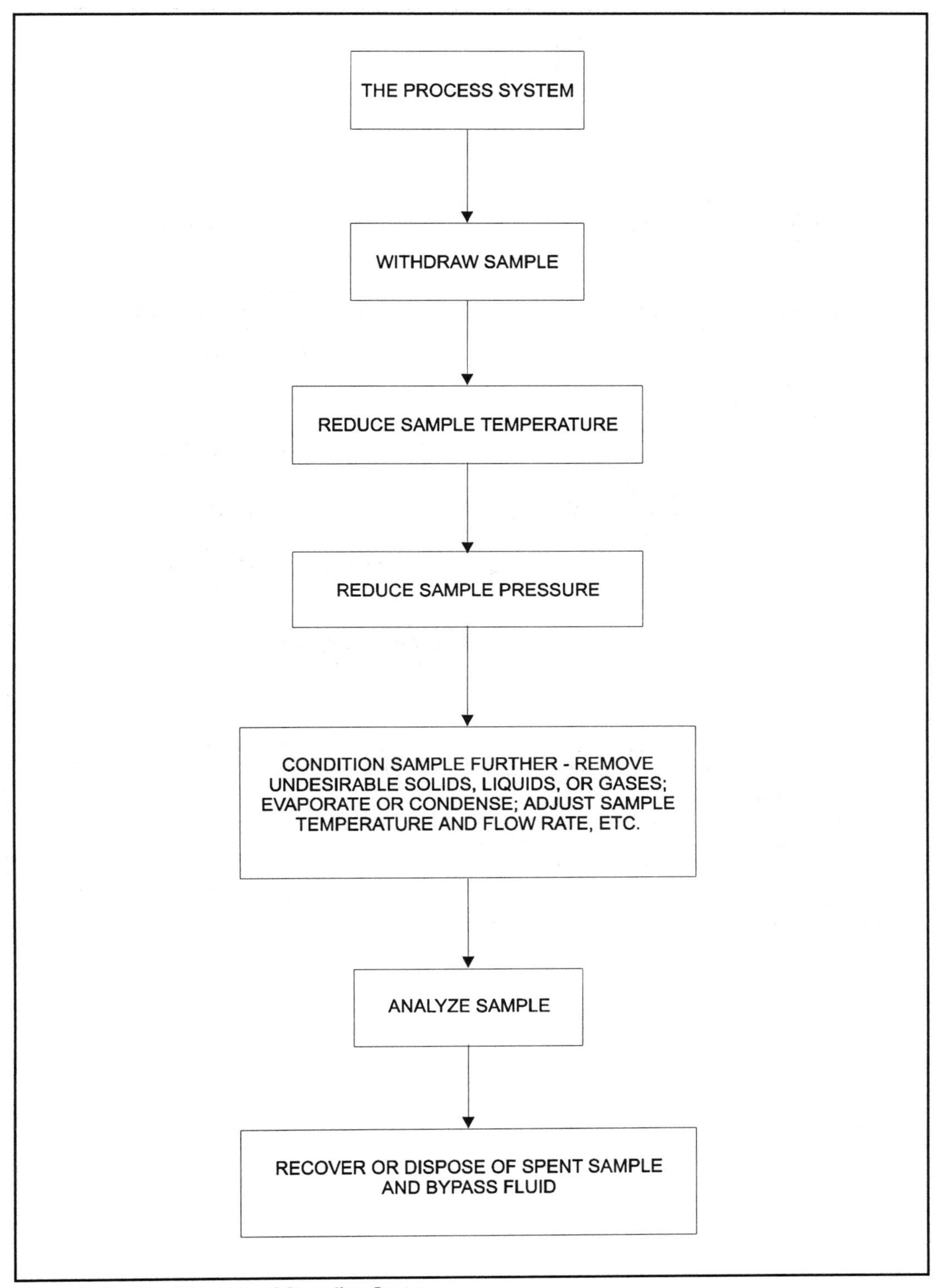

Figure 3-23. Basic Design of Sampling System

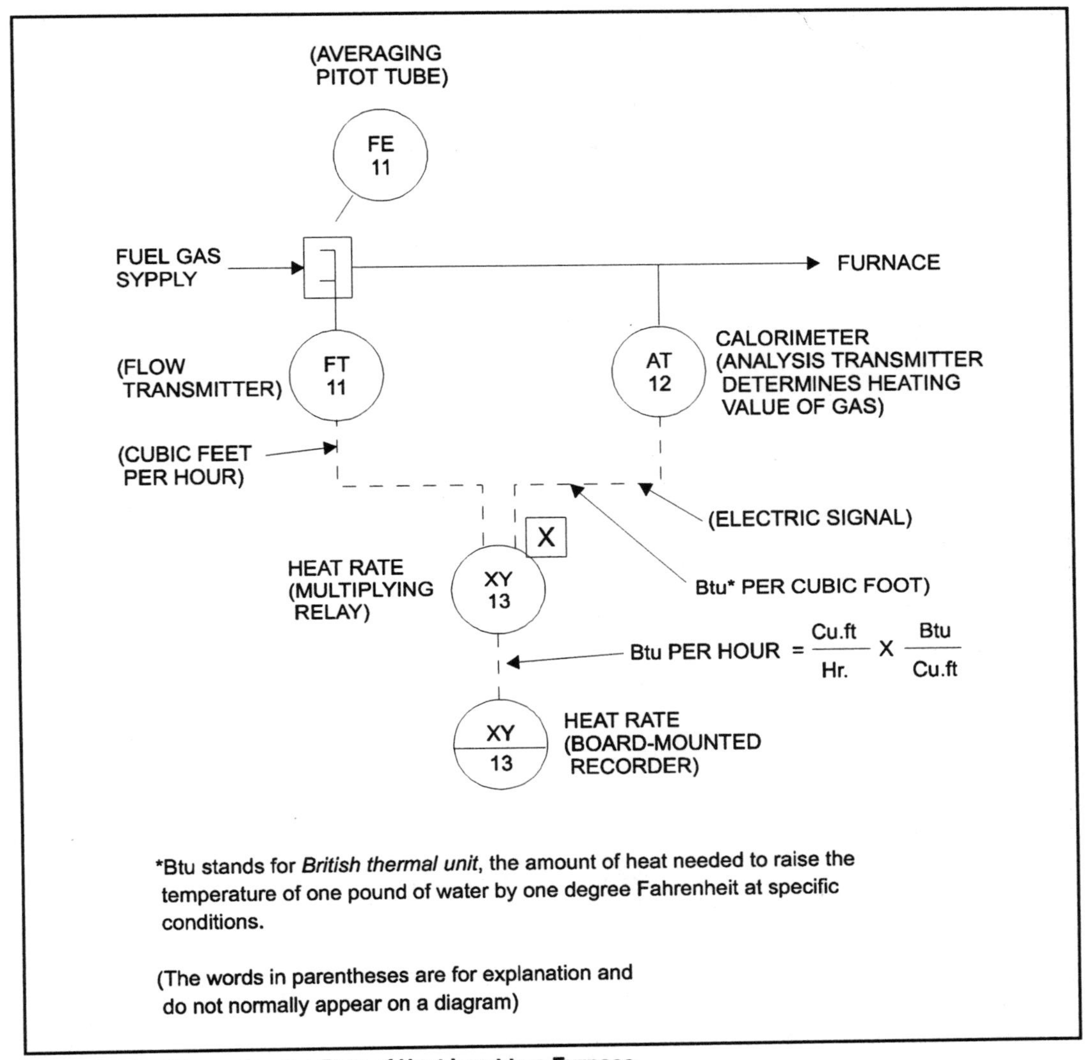

Figure 3-24. Measuring the Rate of Heat Input to a Furnace

4

How Classical Controllers Control

This chapter covers control by controllers. Chapter 7 covers control by logic.

Every process system has a flow of material, energy, or both. The flow of material or energy is manipulated at the command of a controller whose purpose is to keep a process variable at a desired value; this value is termed the *set point*. For example, a pressure controller manipulates a flow of gas, a level controller manipulates a flow of water, and a temperature controller manipulates a flow of electrical energy.

When the controlled system is at steady state, the inflow of material or energy equals the outflow. If the flow balance is disturbed, for example, if more material or energy goes out than comes in, then the controlled variable drifts away from its set point and the controller must call for a corrective action.

4-1 Closed-Loop versus Open-Loop Control

Chapter 2 described how various instruments are connected together to form control loops. Figure 4-1 shows a generalized form of the control loops of Figures 2-1 and 2-2. All these figures illustrate automatic control information traveling around a closed circuit. This arrangement is termed *closed-loop control*.

The set point, which is adjustable, is put into a summing unit, typically by a plant operator. The measurement of actual value of the controlled process variable is also put into the summing unit. This unit subtracts the actual value from the set point, and the resulting output represents the *control error*—the amount by which the value of the controlled variable deviates from what it should be. This error is what the controller tries to eliminate or minimize in closed-loop control.

The summing unit, in most cases, is integral with the controller but is sometimes separate. The measuring elements, which may be separate or combined, are called *feedback elements* because they feed a report on the state of the controlled variable back to the summer so that corrective action may be taken.

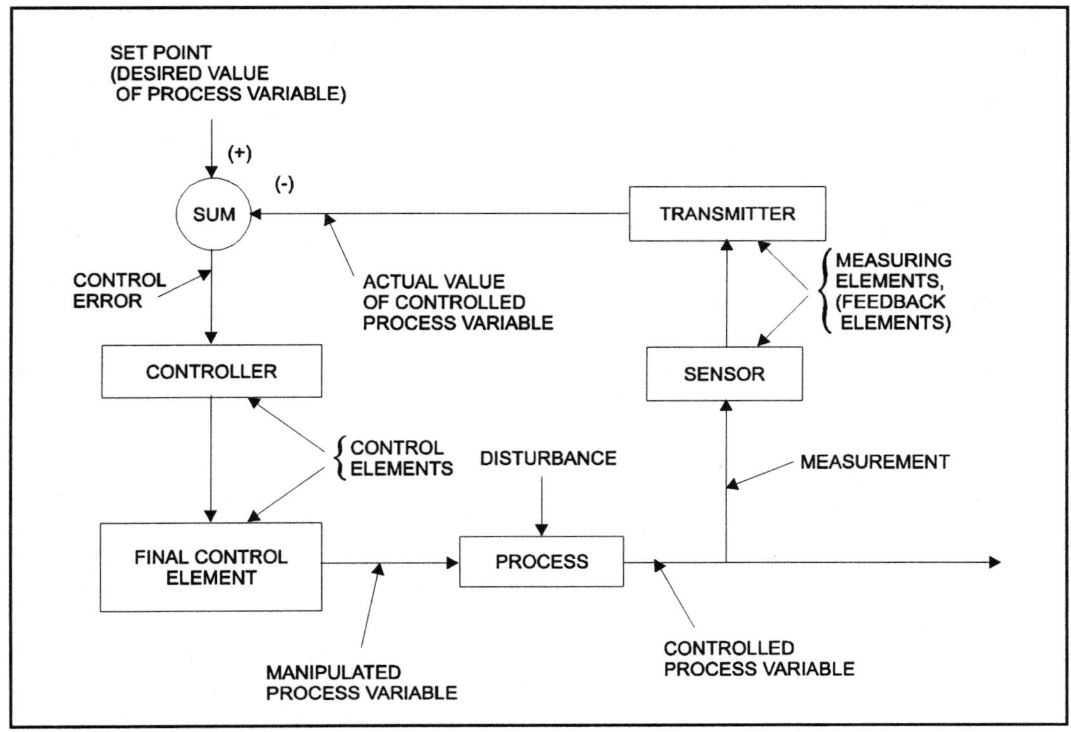

Figure 4-1. Generalized Closed-Loop Control

Most industrial control loops are of the closed-loop type. Another type of control loop is like the closed-loop except that it has no feedback, the control circuit is not closed. This arrangement is known as *open-loop control*, which is shown in generalized form in Figure 4-2.

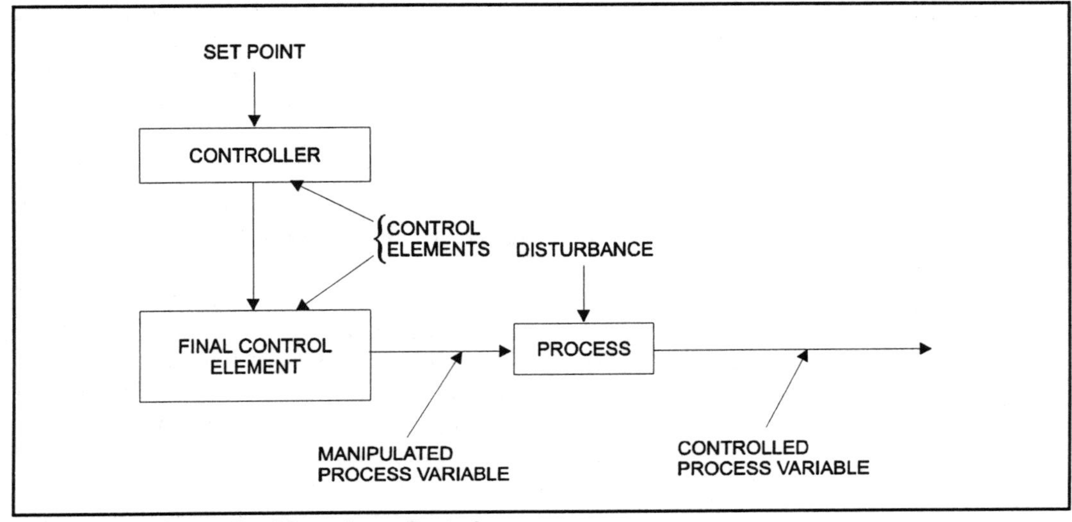

Figure 4-2. Generalized Open-Loop Control

An example of open-loop control is a person driving a car and trying to stay exactly at the highway speed limit. He steps on the gas pedal too hard, then not hard enough, and he frequently readjusts. The driver decides that he has had enough of open-loop control. He switches on the automatic closed-loop, cruise control (see Section 2-1); ah, now he can relax (so far as the speed is concerned).

Similar cases of open-loop control exist in a process plant. Under manual control we may fill a tank, maintain a temperature, match a fluctuating flow, and the like. We have to keep alert. Closed-loop control automatically does these things more reliably, usually better, and it is easier on us.

When open-loop control is used, it is often for manual operation, but it can also be performed automatically. However, it still has no feedback or else it would not be open-loop. An example of this is a chemical feed system for which we wish to use five gallons of reagent B for every twelve gallons of reagent A. We set up an automatic proportioning system to do this, but the system does not check on itself to make sure that the proportion of the reagents is correct.

We may have an alarm to warn us when the mixing is not correct; that would provide information feedback but not automatic control feedback. In such a case, the loop can be said to be closed through the plant operator, who will perform as a feedback control element. But this is not what is meant by "closed-loop control."

Typically, an automatic controller has a built-in manual switch to cut out the automatic control and permit manual control of the output signal.

4-2 How Automatic Controllers Function

For an automatic closed-loop controller to keep a process variable at its set point, it must know whether the variable is at the correct value. But a *Yes* or *No* answer is not enough; the controller must also know, as a minimum, whether the value of the variable is too high or too low. For better control, it must know how much too high or too low the value is, in other words, the magnitude of the control error. For even better control, it may need to know for how long the error existed. For still better control, it may need to know how fast the variable is changing.

These various refinements of control make up different *control modes*, which are described in the next four sections, beginning with *binary control*. A control mode is a specific way in which the input of a controller affects the output.

4-2-1 Binary Control

Binary control—also known as *on-off control* and *two-position control*—is the simplest of all the control modes. It has shortcomings and is the least adaptable of all the types of control, yet it is completely adequate in many cases and is frequently used in process plants. It is also the least costly control and is the type almost always used in home thermostats.

The output of a binary controller is either *On* or *Off*. Its value depends on the following factors:

- The *direction* of the control error.
- The controller action, *direct* or *reverse*.

Automatic controllers have to know whether their measured-variable input has to make a correction up or down. For example, if the temperature is too high, the fuel flow rate must be decreased; if the temperature is too low, the flow rate must be increased.

However, controllers may have an additional kind of operating adjustment to consider. Commonly, an input signal to increase a controller output causes the output to increase. Or conversely, an input signal to decrease the controller output causes the output to decrease. The controller actions are congruent, in agreement, and the controller is described as *direct-acting*. Such, for example, is the case for common house thermostats.

But in the world of industrial process control, it is not uncommon for an input signal to increase a control input and cause the output to *decrease*. Or, an input signal to decrease the controller output may cause the output to *increase*. Here the control actions are in opposition, and the controller is described as *reverse-acting*.

The controller actions, direct- or reverse-acting, are set initially and usually remain unchanged. The summing unit shown in Figure 4-1 determines the direction of the control error, positive or negative.

Section 2-2 described how a binary controller operates in an electric iron. For an industrial process, consider a tank of oil whose temperature is controlled by a binary controller. The oil is heated by steam in a pipe located inside the tank. If the temperature is low, the controller opens a steam valve fully; if the temperature is high the valve closes fully. There is no intermediate flow of steam—the flow is either zero or 100%.

In binary control, the process difference, the operating band, between the *On* and *Off* actions is usually small as, for example, in room temperature control. However, there are cases where the on-off band is intentionally made large in order to minimize the frequency of operation of equipment, such as a pump. This is frequently done for sump level control. A *sump* is a pit into which liquids drain through the force of gravity. When the sump becomes full, a binary level controller starts a pump that operates until the sump is empty. The controller then stops the pump, and the sequence repeats. This kind of binary control is known as *differential-gap control*.

Binary control is simply a switching action, and the same instrument can be used for any other switching operation, for example, to operate an alarm. Then the instrument is tagged as a *switch*, not as a controller.

Almost all control systems operate with varying process loads. The *process load* is the amount of material or energy that must be manipulated in order to control the controlled variable. The load depends on whether the system is operating at capacity or less than capacity, on the weather, on the quality of the plant feed, and on other factors.

Binary control is theoretically and actually incapable of keeping the controlled variable within a given operating band for more than one process load. At other loads, the band moves up or down, depending on whether the load has increased or decreased. The shift of the control band is called *offset* or *droop*, which is small or large depending on the size of the load change.

The offset is a reason why a thermostatically controlled house that is poorly insulated requires that the set point be raised manually to keep the house comfortable when the weather turns colder. This thermostat is a binary controller whose operating band has fallen because the house heating load has increased.

4-2-2 Proportional Control

Proportional control, also known as *single-mode control*, provides a modulated output that can have any value from the bottom to the top of the output range. To *modulate* is to *adjust* or *regulate*. The output depends on the following factors:

- The direction and *magnitude* of the control error.
- The *gain*, or sensitivity, of the controller.
- The controller action, direct or reverse.

Controllers and processes, like people, are sensitive to a greater or lesser degree. If we say the wrong thing to some people, they flare up in anger; we consider them very sensitive. Say the same thing to other people and they may laugh it off or ignore it; they are less sensitive. The greater the sensitivity, the greater the reaction to an input stimulus.

Sensitivity is normally expressed in terms of *proportional gain* or often simply *gain*, which is equivalent to the amount of reaction divided by the amount of stimulus. *Gain* is defined as the change of output corresponding to a given change of input divided by the change of input. For most controllers, the proportional gain is adjustable. (Instead of the preferred term, proportional gain, the opposite term, *proportional band*, is sometimes used. Proportional band is the input change divided by the output change and is also usually stated as a percentage.)

A controller whose output changes by 10% of its initial value when the input changes by 8% of its initial value has a gain of 1.25, or 125%. If the same input change causes an output change of 4% of its initial value, the controller gain is 0.50, or 50%.

The word *gain* applies also to instruments other than controllers and to complete loops. The input to a control valve typically comes from a controller. If the valve input signal changes from 5 psig to 7 psig (an increase of 40%) and causes the output (which is process flow) to change 20% from 100 gpm to 120 gpm, the valve gain is 0.50, or 50%, or 10 gpm per psi.

The same holds true for processes: If the steam input to a water heater changes from 10,000 pounds per hour (pph) to 12,000 pounds per hour, causing the water temperature to change from 180°F to 200°F, the heater gain is 0.010, or 1.0%, or 1°F per 100 pounds per hour.

Now replace the binary controller of Section 4-2-1 with a proportional controller. If the oil temperature is a little too low, the controller makes the steam valve open a little; if too low by a large amount, the valve opens farther. If the temperature is a little too high, the controller makes the valve close a little; if too high by a large amount, the valve closes farther. Under normal conditions, the controller output is modulated, the control valve modulates the steam flow, and there is generally some flow.

Like a binary controller, a proportional controller is subject to offset and is incapable of keeping the controlled variable at its set point at more than one process load. The offset is the reason why a thermostatically controlled automobile cooling system or a house runs cooler in cold weather than in hot weather. In this case, the thermostat is a self-actuated regulator that provides only proportional control.

4-2-3 Proportional-plus-Integral Control

Proportional-plus-integral control is also known as *two-mode* control, *PI control,* and *automatic-reset* control. The integral mode is also called *floating control.* Just as for proportional control, the controller output is modulated, but the value of the output depends on the following factors:

- The direction, magnitude, and *duration* of the control error.
- The gain of the controller, which depends on the proportional gain and the *integral time*, both of which are adjustable.
- The controller action, direct or reverse.

Now let us use a PI controller for the oil temperature. This controller performs just as the proportional controller does, but it has an added feature: Assume the oil temperature is low. The proportional mode output calls for an increase in steam flow, which is supposed to bring the temperature up to the set point. If the temperature remains low, the integral mode gradually adds to the output, and the valve opens farther. So long as a control error remains, the integral mode keeps adding to the output, the valve keeps opening, and the temperature finally reaches the set point. The control error is now zero, so the controller is satisfied; its output remains constant, and the steam flow remains constant until the oil temperature is disturbed again. Then the control system again responds to the deviation from set point. The additive action of the integral mode reinforces the action of the proportional mode in either direction, up or down.

Because of the integral mode this controller does not have control offset. At any new steady load, the controller returns the controlled variable to its set point. (Compare with Sections 4-2-1 and 4-2-2.)

4-2-4 Proportional-plus-Integral-plus-Derivative Control

Proportional-plus-integral-plus-derivative control is also called *three-mode control* and *PID control*. The derivative mode is also called *rate action*. A three-mode controller modulates its output, whose value depends on the following factors:

- The direction, magnitude, duration, and *rate of change* of the control error.
- The gain of the controller, which depends on the proportional gain, the integral time, and the *derivative gain*, all of which are adjustable.
- The controller action, direct or reverse.

A PID controller used for the example of oil temperature control in the previous section adds the following feature to those of the PI controller: the derivative mode changes the controller output according to how fast the control error is changing. If the error is not changing, the derivative mode does nothing. If the error changes slowly, the derivative mode changes the output a little during the error change. If the error changes rapidly, the derivative mode makes a large change to the output during the error change. The purpose of the derivative mode is to give a boost to the corrective action of the controller, especially when the controlled variable is changing fast. In this way, the derivative action tends to prevent the error from becoming excessive before the proportional and integral modes can regain the upper hand. The derivative mode is used mostly for slow systems.

To control our oil temperature, let us use a three-mode controller. If the temperature fall below the set point is slow, the controller acts very much like a two-mode controller. If the fall is rapid, the action to increase the opening of the steam valve comes mainly from the proportional and derivative modes. This reduces the rate of fall and causes the derivative action to become less important. In the meantime, because of the prolonged deviation from set point, the integral action, which responds to the duration of the deviation, becomes more important. Finally, the temperature is brought back to the set point and stays there, the derivative effect has dropped to zero, and the controller output and the steam flow hold steady at the set point.

4-2-5 PROCESS CONTROLLABILITY

The four control modes we have discussed—binary, proportional, integral, and derivative—are the ones commonly used. Their responses are summarized in Table 4-1.

Table 4-1. Controller Responses to Errors

Control Mode		Responds to the Control Error			
		Direction	Magnitude	Duration	Instantaneous Rate of Change
Binary (on-off)		●			
Modulating	Proportional (P)	●	●		
Modulating	Integral (I)	●	●	●	
Modulating	Derivative (D)	●	●		●
Modulating	PI	●	●	●	
Modulating	PD	●	●		●
Modulating	PID	●	●	●	●

The choices of binary control or modulating control, and which type of modulating control to use, depend on how difficult it is to control the process. The easier a process is to control, the simpler the controller can be. In fact, a process that changes very infrequently and is otherwise easy to control may be handled satisfactorily by manual control, assuming that the possibility of misoperation or nonoperation because of human failure is taken into account. On the other hand, the more difficult the process is, the more sophisticated the controller and its adjustments generally must be.

Factors that affect the ease of control are stated in Table 4-2, and are explained in Section 4-2-5-1. Process factors that influence process controllability are items 1 through 10. Instrument factors are items 11 through 13.

Table 4-2. Factors Affecting Process Controllability

Increasing Process Factor	Effect on Ease of Control	
1. Permissible deviation from set point	Helps	
2. Load-change magnitude		Hurts
3. Load-change rate		Hurts
4. Process nonlinearity		Hurts
5. Resistance		Hurts
6. Capacitance	Helps	
7. Dead time		Hurts
8. Process noise		Hurts
9. Environmental change		Hurts
10. Control valve pressure drop	Helps	
Increasing Instrument Factor	**Effect on Ease of Control**	
11. Measurement nonlinearity		Hurts (in most cases)
12. Incorrect valve flow characteristic		Hurts
13. Signal noise		Hurts

4-2-5-1 Process Control Factors

1. *Permissible Deviation from Set Point*. Close control of the controlled variable is sometimes important, sometimes not. The looser the requirement, the greater the acceptable deviation from the set point and the easier is the control. Particularly in level control, a wide deviation of the level is often planned to enable the tank to absorb uncontrolled flow surges into or out of the tank without causing the manipulated flow to surge correspondingly; this type of control is called *averaging control*.

2. *Load-Change Magnitude*. A process has a flow of material or energy. The larger the range of changes of the flow between minimum and maximum, the more difficult is the control.

3. *Load-Change Rate*. If the flow of material or energy changes rapidly, the control is more difficult.

4. *Process Nonlinearity*. For a discussion of this, see Section 6-1-2 covering control valve flow characteristics.

5. *Resistance*. This is the characteristic of impeding the flow of material or energy. All materials permit heat to pass through them, but not equally well. Some materials, such as building bricks or a wool coat, conduct heat poorly; they have high resistance and are called *thermal insulators*. Other materials, such as a metal spoon or a copper wire, conduct heat well; they have low resistance and are good *thermal conductors*. The thermal resistance depends on the type and the amount of material in the path of the energy flow.

There are insulators and conductors for electricity as well.

The flow of materials may also be inhibited by a resistance. Just as a narrow street impedes the flow of traffic, so a small-diameter pipe is more of a hindrance to liquid flow than a large-diameter pipe. (The flow resistance of pipe is related to item 10, control valve pressure drop.)

The greater the resistance, the more difficult it is to control a process because the resistance tends to inhibit or distort corrective control actions.

6. *Capacitance.* This is a property of storing material or energy. It is defined as the change in the amount of material or energy needed to make a unit change in a process variable. For example, capacitance is the number of gallons of water required to change a tank level by one foot. In other words, to make a change in a controlled variable, some quantity of a manipulated variable must be supplied or removed; that quantity divided by the change is the capacitance.

   Capacity is the maximum quantity of material or energy that can be stored in a device or system.

   Figure 4-3 shows two tanks having the same capacity but different capacitances. So far as level control is concerned, the process is more sensitive in the tall tank than it is in the short tank because the capacitance of the tall tank is smaller; for the same change of volume in both tanks, the level in the tall tank will have a larger change. It has a faster response but is more difficult to control within a given range. Increasing the capacitance for a given control range improves the stability of the controlled variable in response to an upset.

   Capacities and capacitances for typical process variables are summarized in Table 4-3.

Table 4-3. Units for Typical Process Capacities and Capacitances

Controlled Variable	Capacity	Capacitance
Chemical analysis	Pounds	Pounds/percent composition
Liquid level	Gallons	Gallons/foot
Pressure	Cubic feet	Cubic feet/psi
Thermal	Btu	Btu/degree

7. *Dead Time.* This is a period of delay between two related and sequential actions, such as the beginning of a change of input and the beginning of a resulting change of output. Dead time may be called *transportation lag* or *delay time.*

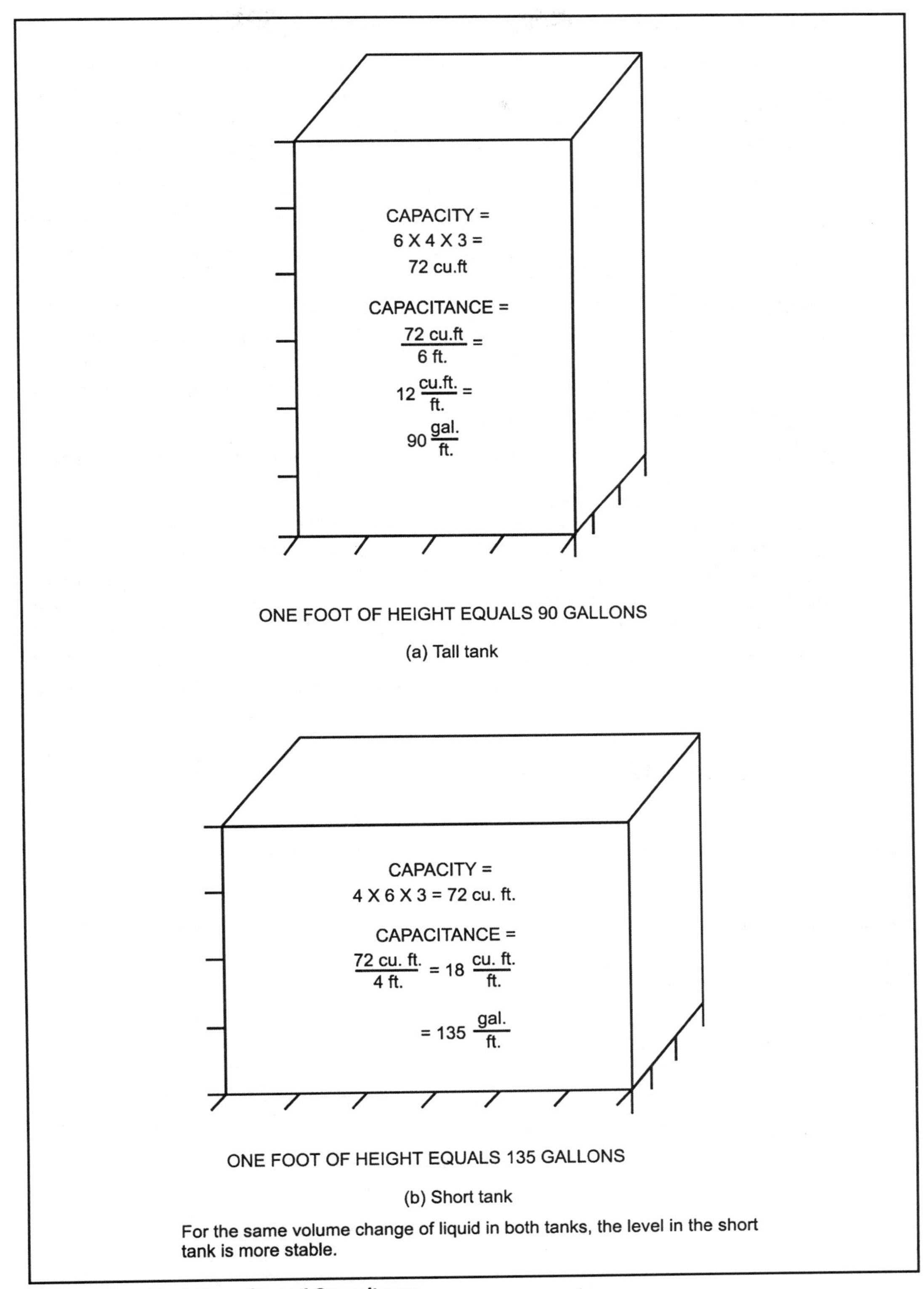

Figure 4-3. Tank Capacity and Capacitance

Suppose a person wants to take a shower. She stands aside and opens the hot-water faucet. Water comes out immediately, but, unfortunately, it is very cold. Six seconds later, the water starts to become warm, then hot. She adjusts the temperature and steps under the shower.

Why does the person have to wait so long for warm water? The reason is because the shower is at the opposite corner of the house from where the hot-water tank is. It takes six seconds for the hot water to travel from the tank to the shower. These six seconds are the dead time.

Similar situations occur in plant processes where the results of an upset or operating change must first be sensed, and then a flow of material or energy must be changed accordingly to achieve a control result. Dead time in any part of the control loop degrades the control because corrective action cannot fully begin during the dead-time period. In the meantime, the control error has been growing.

In most cases, the dead time is not large and is not really troublesome. In some cases, especially for analysis control, the dead time can create significant difficulties that require sophisticated control schemes to overcome them.

8. *Process Noise*. *Noise* is a persistent disturbance that obscures or reduces the clarity or quality of a measurement. If you and I are conversing where there is a lot of audible noise, we find it difficult to understand each other's message.

   Similarly, there can be a noisy flow or level that we wish to measure. The process noise is created by fluid turbulence, high velocity, eddy currents, or waves, which cause a rough and bumpy measurement even when the flow rate is steady. (There may also be audible noise, but this does not affect the measurement, except possibly for certain process instruments that sense sound.) The greater the process noise, the more difficult the control.

   To improve the control of noisy processes by quieting the noise, a fixed resistance or other suppression device may be added in the sensing line or in the measuring circuit.

9. *Environmental Change*. Changes from a sunny sky to a cloudy sky, day to night, and winter to summer affect the control of some processes because of the changes they cause, such as cooling of water temperature or thermal effects on outdoor process equipment. Changes in the wind or atmospheric humidity affect the performance of equipment that cools water by evaporation.

   Changes of barometric pressure affect gage-pressure instruments and thus affect operating values, especially at very low process pressures.

   Such changes affect the process load, with the control effects noted in items 2 and 3 of this list. The disturbance that is shown affecting a process in Figure 4-1 may be a change in the environment.

10. *Control Valve Pressure Drop*. For a discussion of this with respect to controllability, see Section 6-1-2.

4-2-5-2 INSTRUMENT FACTORS

1. *Measurement Nonlinearity.* The gain, or output-to-input relationship, of a sensor may or may not be linear, as described in Section 3-5-2-1 for "Differential Pressure." Nonlinearity is caused by the fundamental relationship of a process variable to its sensor output and is separate from the minor imperfections of an instrument. A signal converter can be added to a loop to change a signal from nonlinear to linear.

   Figure 4-4 illustrates input/output (I/O) relationships for linear and nonlinear sensors.

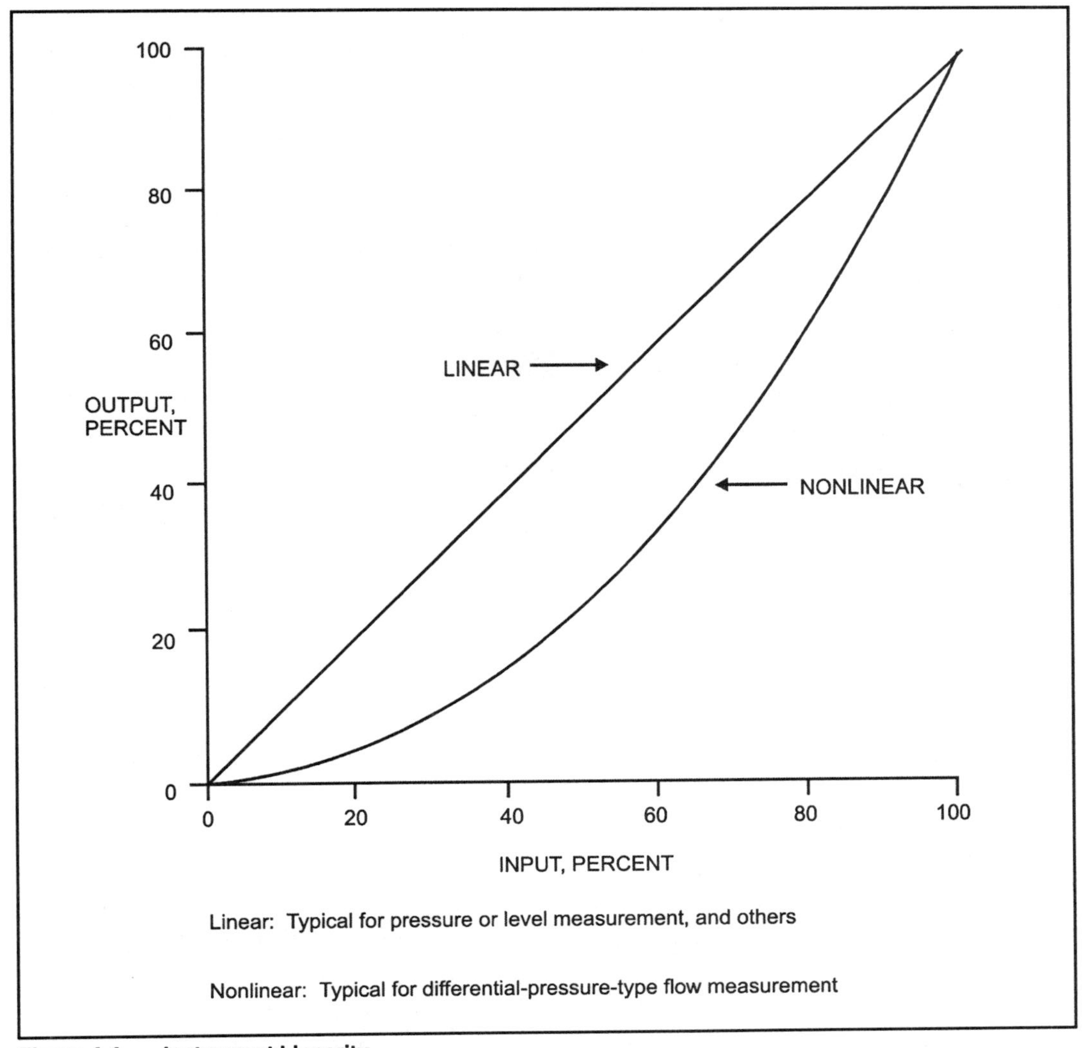

Figure 4-4. Instrument Linearity

2. *Incorrect Valve Characteristic*. For a discussion of this, see Section 6-1-2.
3. *Signal Noise*. Signal noise, which is particularly prevalent for electrical signals, degrades the quality of the information carried by the signal, thereby degrading the effectiveness of the control loop. Electrical noise may be caused by interference from other electrical signals, power lines, and electrical machines. Everyday examples of signal noise are radio static from the atmosphere or visual interference on a TV screen caused by a fluorescent lamp, an electric motor, or a nearby amateur radio broadcast.

   To cope with electrical noise, precautions should be taken with the instrument wiring. These precautions include the use of shielding and grounding, twisted wire pairs, separation from potential sources of trouble, and proper terminations for the wiring.

4-2-6 Controller Tuning

A modulating controller must be tuned for its specific service. *Tuning* is the procedure of adjusting the controller for optimum performance. This may involve adjusting the sensitivity of each of the controller's control modes as well as using auxiliary dynamic elements, if any (see the discussion of lead-lag relay in Section 4-3-5). There are mathematical procedures and process studies that can be used to estimate the best preliminary tuning adjustments for a given controller. In addition, the controllers are adjusted in the field by trial and error and by experience. Even when sophisticated methods are used, the resulting tuning should be confirmed by field trial.

Off-the-shelf electronic controllers are available that have the ability to tune themselves automatically.

4-3 Combination Control Systems

The basic types of control loops shown in Figures 4-1 and 4-2 are used in several important combinations, as follows:

- Ratio control.
- Cascade control.
- Selective control.
- Split-range control.
- Feedforward control.

Each of these combinations can be used in conjunction with one or more of the other combinations.

4-3-1 Ratio Control

Ratio control is control of the ratio of two process variables. One variable fluctuates according to requirements of the process and is called the *wild variable*. The other variable is proportioned to the wild variable and is called the *manipulated variable*. A process such as blending gasoline uses ratio control to proportion a number of components, using different ratios for summer gasoline and winter gasoline.

Figure 4-5 shows how chemical A is blended with chemical B in a desired proportion. A is the wild stream, B the manipulated stream. The two flows are measured by two transmitters that inform a ratio-computing relay what the instantaneous flow rates are. The relay divides a pneumatic signal, for example, representing flow A by the signal representing flow B, and sends the resulting ratio signal to a controller. The internal set point of the controller has been adjusted to call for the B flow to be 5/12 as much as the A flow. The controller continuously adjusts the flow of B so that the blend of A-plus-B contains five parts of B for every twelve parts of A, no matter how the A flow varies.

4-3-2 Cascade Control

Cascade control is control in which the output of one controller establishes the set point for another controller. Consider an oil heater that uses steam in a heating coil to heat a stream of oil, as shown in Figure 4-6(a). This system would provide good control of the oil temperature if the supply pressure of the heating steam, the oil flow, and the oil inlet temperature were all constant. Then the temperature controller would have the control valve pass exactly the correct amount of steam to keep the oil outlet temperature at the set point, and thermal equilibrium would exist in the heater.

However, the steam supply pressure fluctuates widely because of a fluctuating use of steam by other users. The rate of heat transfer from the steam to the oil depends on the difference between the steam temperature and the oil temperature. The steam temperature depends on the steam pressure in the heater; the higher the pressure, the higher the temperature. When the steam supply pressure rises or falls, the immediate effect is to change the steam pressure inside the heater. This increases or decreases the steam-to-oil temperature difference, and the oil correspondingly becomes too hot or too cold. No correction is made until the temperature controller finally begins to react. In the meantime, there has been a temperature control error. Temperature control is usually slow, so the error may become large and may last for a long time.

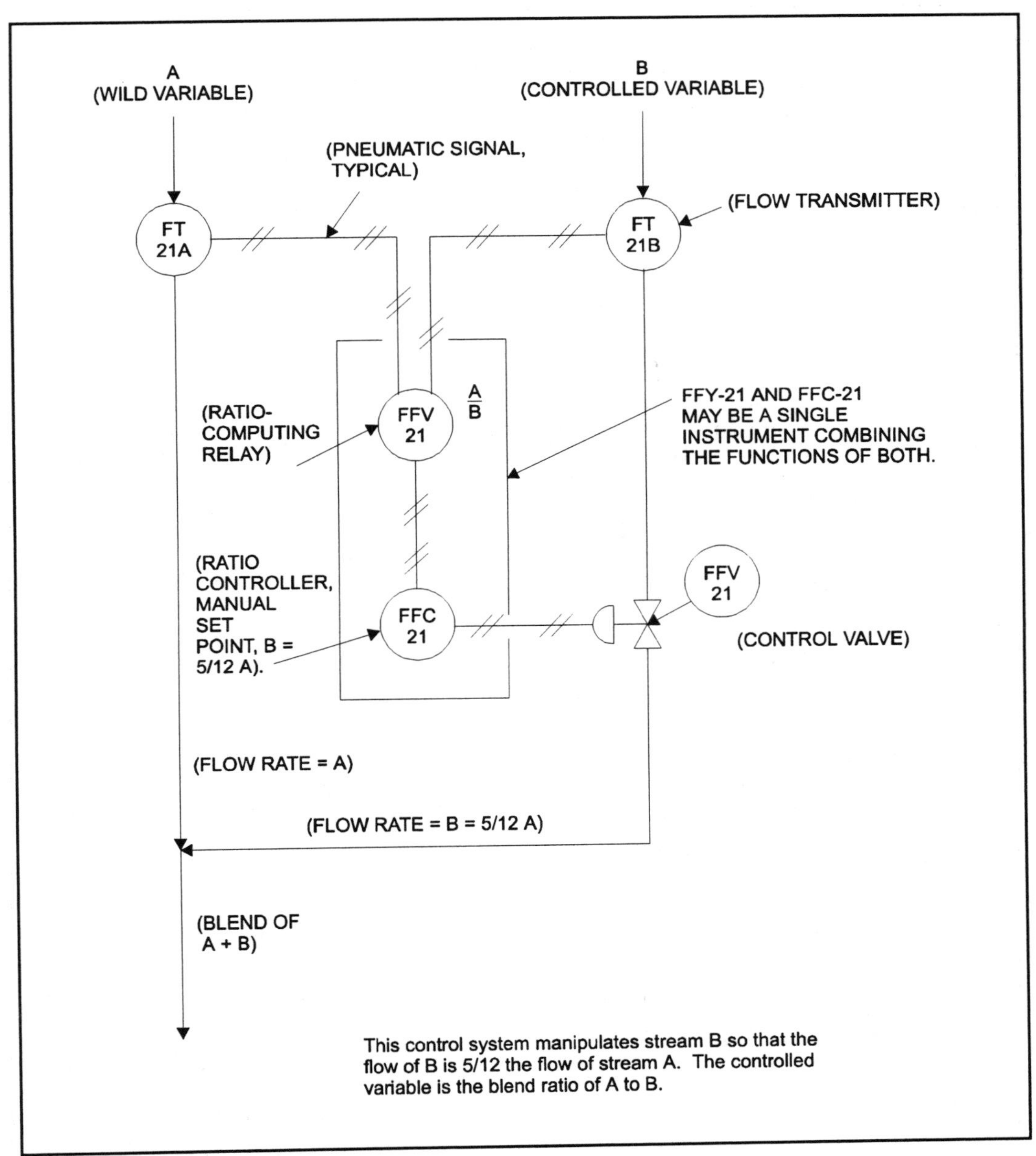

Figure 4-5. Flow-Ratio Control

The temperature control can be improved greatly by the system shown in Figure 4-6(b). When the steam supply pressure shifts up or down and causes the pressure to change inside the heater, the pressure controller senses the heater change and readjusts the control valve to bring the heater pressure back to its set point at that time. The pressure control loop acts fast, and it restores the thermal equilibrium with only a slight and brief disturbance to the oil temperature. Thus, the effect of steam pressure fluctuations is taken care of nicely for a given heater load.

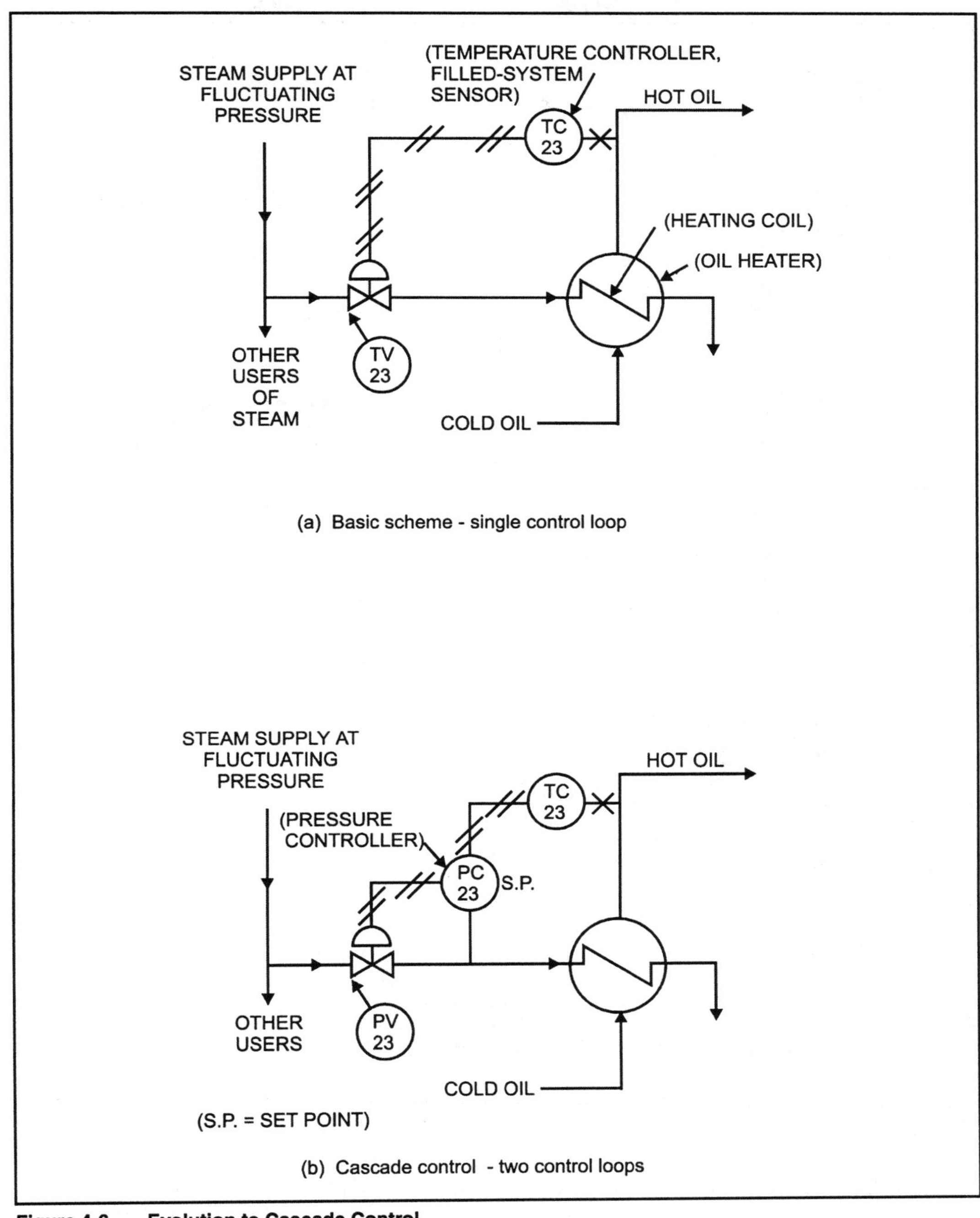

Figure 4-6. Evolution to Cascade Control

But what happens if the heater load, the heat transfer requirement, changes? The load depends on the oil flow and its inlet temperature. The temperature controller feels the change in the hot oil temperature and tells the pressure controller to hold a new steam pressure. The pressure controller is provided with a set-point element that accepts a command signal. The pressure controller quickly does as it is told to provide the new steam pressure, and it continues to act promptly all by itself to correct the pressure whenever it fluctuates. However, if the load changes are severe and the control requirements are tight, additional measures may be required, as described in Section 4-3-5.

The cascade control system has two feedback controllers but only one final control element. In the example, the temperature controller typifies a *primary controller* or *master controller*. The pressure controller typifies a *secondary controller* or *slave controller*. The entire secondary control loop for pressure control can be looked at simply as an elaborate final control element for the temperature controller.

A general requirement for all cascade control systems is that the secondary control loop be appreciably faster than the primary control loop. If the secondary is not faster, it tends to destabilize the primary control loop instead of stabilizing it; a continuing, undesirable oscillation, known as *hunting*, may occur.

4-3-3 Selective Control

Selective control uses two controllers and only one control valve or other final control element, just as cascade control does, but with a big difference. Cascade control has one controller that controls the other. In selective control, each controller is independent of the other. Either controller may be in control at any given time, but when one controller is in control, the other is overridden and is ineffective. The controller that is calling for the greater corrective action in a predefined safe direction is then in charge; the other controller is twiddling its thumbs, figuratively speaking. The act of selecting between the two controllers is also known as *auctioneering*.

The concept of selective control is clarified by the example in Figure 4-7, which shows a tank whose level is controlled by modulating a control valve in the drain line. The flow rate of the tank drain is controlled by using the same valve. There are two limiting requirements: the level should not get too low, and the flow should not get too high. When the level is low, the level controller is normally in control and cuts back the outflow. When the flow is high, the flow controller is normally in control and cuts back the outflow. At all times, the valve takes the less-open position of the commands from the two controllers.

The choice of which controller shall override the other is made automatically by a selector relay, which makes a smooth transition from one input signal to the other. The function of this separate relay may be built into a controller.

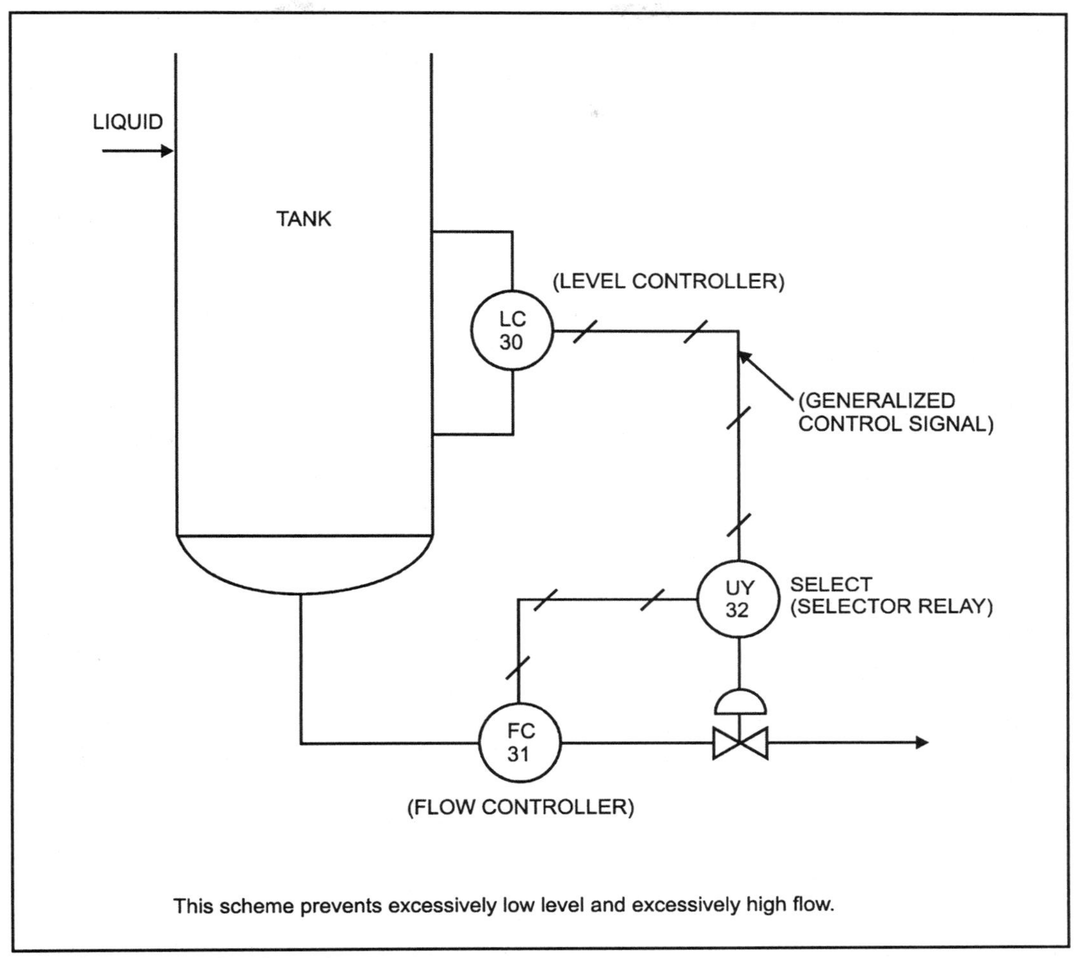

Figure 4-7. Selective Control

4-3-4 Split-Range Control

There are control loops that use one controller and two control valves, both of which modulate. Figure 4-8(a) shows a temperature control scheme for a batch process using a chemical reaction tank that requires the reaction temperature to be constant. To start the reaction, the tank must be heated; this requires a flow of steam through a heat-transfer coil. Later the reaction produces heat and the tank must be cooled; this requires a flow of cooling water through another coil.

Smooth control of the temperature is provided by the following basic system: the output of the temperature controller changes gradually as the tank temperature increases. When the controller calls for the heating valve to be wide open, the cooling valve is fully closed. When the cooling valve is wide open, the heating valve is fully closed. Halfway between, both valves are closed so there is neither heating nor cooling. Each valve fully strokes in reverse to and in sequence with the operation of the other valve. This valve arrangement is known as *split-ranging* and is illustrated in Figure 4-8(b).

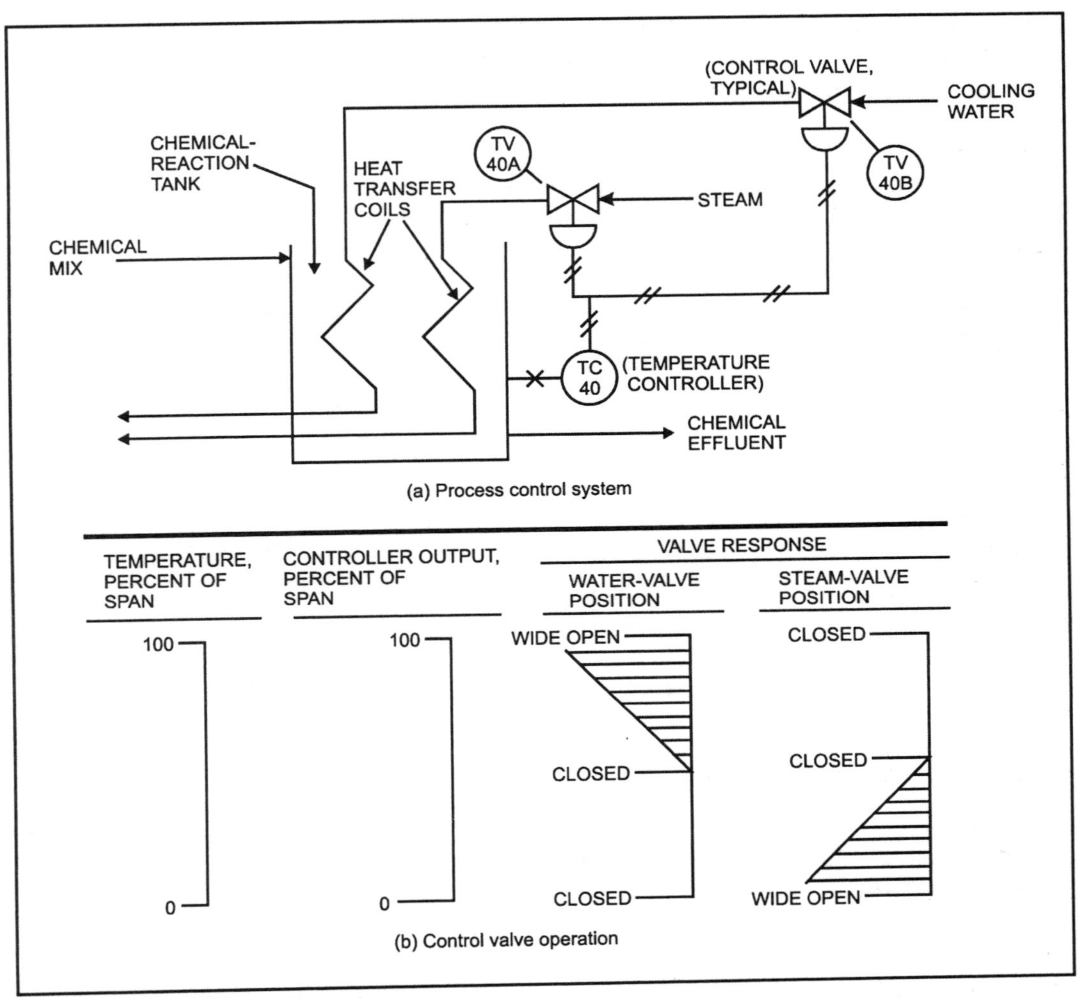

Figure 4-8. Split-Range Control

4-3-5 Feedforward Control

Let us look back at the oil-heating process of Figure 4-6(a), showing a simple feedback loop for controlling the hot-oil temperature. Assume that there will be large variations in (a) the flow of oil to the heater, and (b) the temperature of the entering cold oil. Each of these changes will cause sudden and large changes in the heating load, and the flow of steam must change correspondingly. The temperature controller will eventually bring the system to a new equilibrium state. But temperature readjustment is relatively slow, and, in the meantime, the exiting oil will be too hot or too cold. The feedback control system is not adequate.

Feedforward control can give excellent control during these upsets. Feedforward control makes corrections to minimize, or theoretically to eliminate, the error of the controlled variable strictly on the basis of information concerning factors that would upset this variable if the corrections were not made. Upsetting factors to be taken into account are changes to properties of process feed flows and other conditions that may affect plant operation.

For example, we are driving a car at thirty-five miles per hour on a level road. We are approaching a hill with a steep upgrade, but we wish to keep our speed constant. We step on the gas pedal the moment we reach the incline in order to increase the power output of the engine. In effect, we evaluate the speed upset that would occur if we were to make no operating change when we begin climbing the hill, and we increase the fuel flow at the right time by the right amount to avoid an upset to the speed. This example uses crude manual control, yet it illustrates the principle involved.

To apply feedforward control to our oil system, the control loop must calculate the steam flow—the amount of heat—required to match the cold-oil requirement at every moment, and then it must provide exactly that much steam. To make this calculation, a mathematical model—an *algorithm*—must be developed to simulate the performance of the oil-heating system. The model will need the following information:

- Oil flow rate (by measurement).
- Cold-oil temperature (by measurement).
- Specific heat of the oil (a thermal property, available from reference books).
- Heat of vaporization of water at the pressure inside the steam coil (a thermal property, available from reference books). (*Heat of vaporization* is the heat required to boil the water at a given pressure.)

All this information is put into a small calculating system, UY-25, which may be made up of standard commercial instruments. The control system is shown in Figure 4-9(a). UY-25 is informed promptly when the oil flow rate or the cold-oil temperature changes. The calculator does its arithmetic, determines what the steam flow should be for the new conditions, and control valve UV-25 changes the steam flow accordingly.

Though feedforward control can improve performance greatly, it has the following weaknesses:

- We cannot make a perfect mathematical model of the process and its equipment.

- Every instrument, including control valves, performs imperfectly and contributes to the functional inaccuracy of the control system.

- The design performance of the instruments, and also of the process equipment, is further degraded by environmental changes, aging, corrosion, and other factors (see Chapter 9).

The feedforward errors become intolerable if they are allowed to accumulate. However, we know that feedback control is capable of eliminating errors after they occur. Figure 4-9(b) shows temperature controller TC-29 added to provide feedback control to supplement the feedforward control. When there is a load change, feedforward will greatly reduce the initial temperature error, and the remaining error will be eliminated by feedback. This will reduce both the magnitude of the control deviations and their duration. Such use of feedforward and feedback typically ensures that the controlled variable has a minimum upset and ends up at its correct value.

The control system now includes a steam flow controller, FC-28, which provides flow feedback solely to increase the accuracy of the flow control. The scheme of Figure 4-9(a), lacking the flow controller, has open-loop control of the steam flow, which may be quite inaccurate because of control valve inaccuracies; the inaccuracy is often significant for the slow temperature system.

The system also includes a dynamic element, FY-26, a *lead-lag relay,* whose purpose is to compensate for the difference between the speed of a change in the oil flow and the speed of the resulting change in heat flow from the steam. FY-26 works by speeding up or slowing down signal changes to balance the process responses; its effect is adjustable and is part of the tuning operation for the system.

Feedforward systems are used mainly for difficult-to-control processes, which are usually slow processes with much dead time. Feedforward has worked wonders in improving some processes, such as distillation, electric power generation, and chemical manufacturing.

In theory, feedforward control is capable of absolutely perfect control—no control error. In contrast, feedback control is theoretically *incapable* of perfect control because its corrective action begins only after a control error already exists. Paradoxically, the theoretically inferior control by feedback predominates in actual practice because it is adequate in most cases and is less complex and less expensive.

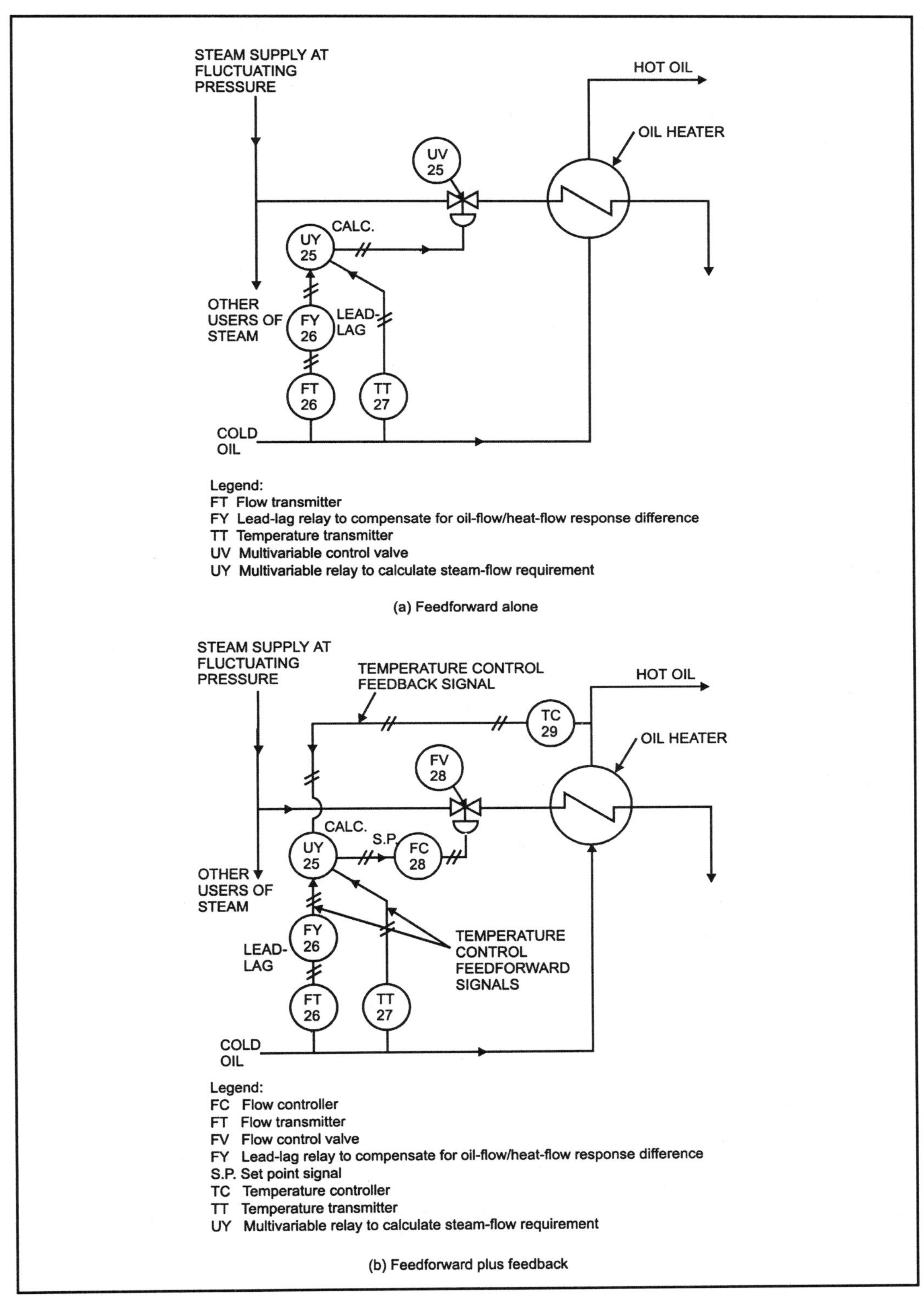

Figure 4-9. Feedforward Temperature Control

How Advanced-System Controllers Control

5-1 How Fuzzy-Logic Controllers Control

5-1-1 Condensed History

The logic system of either *Yes* or *No* was developed by the ancient Greek philosopher Aristotle following work by Pythagoras. This system is called *Aristotelian logic,* also *binary logic,* and it offers *two* choices, no more and no less. The roots of vagueness also began with the Greeks: for example, we have Zeno's Paradox: How many individual grains of sand can be removed from a sandpile before it is no longer a pile?

Albert Einstein said, "As far as the laws of mathematics refer to reality, they are not certain, and as far as they are certain, they do not refer to reality."

Fuzzy logic was invented in the United States by Lofti Zadeh, who made it public in 1965. He understood that many ideas are defined better by words than by mathematics, and he recognized that human intelligence and language can be linked by mathematics.

5-1-2 What Is Truth?

There are two kinds of logic used in process control: the classical or traditional and conventional logic now known as *crisp logic* in the context of the fuzzy-logic world, and the relatively new fuzzy logic. The word *crisp* means precise or exact, the opposite of *fuzzy,* which means vague to some extent.

A crisp-logic system calls for a process state or element to be either *On* or *Off, True* or *False, Yes* or *No*. In numerical terms, the process is in state 1 or 0. By contrast, fuzzy logic (a very apt term) recognizes the states 1 and 0, but, very importantly, it also accepts all numerical values between 1 and 0. For example, a fuzzy valve may be half open, in state 0.5, or open in the amount of 0.84 or 0.15. All these numbers, except 0, represent some truth; 1 is total truth, 0 is total lack of truth.

A fuzzy-logic control application operates according to rules that are based on linguistic terms such as *small, moderate,* and *large*. Each term covers a zone of truth. Such zones indicate the extent of their deviation from a desired or true value. The control goal is zero deviation. The processing system for fuzzy logic is a *fuzzy network* (see Figure 5-1).

The preceding are generalities that apply to fuzzy logic. However, fuzzy logic applies to very disparate control situations, and the details of fuzzy logic tend to be different for each case. Therefore, the examples in the next section are merely representative.

5-1-3 Fuzzy Logic

Problem: Try balancing a stick on a fingertip; this works poorly at best because of instability and manual overreactions. Now try balancing an inverted pendulum weight between two electromagnets; this works well and is stable. Fuzzy logic can stabilize the pendulum motion in a few seconds.

A fuzzy network may be used with a conventional process controller that is off its set point by a small amount, not a moderate or large amount. The network reacts by correcting the small amount of error. If the set point were in error by a moderate or large amount, an equivalent correction would be made. Fuzzy logic minimizes the over- and undershooting of set points.

There is some overlap of adjacent zones in the network. Despite the fact that system inputs cross zone boundaries, fuzzy systems produce smooth and continuous outputs, and the answers they produce are crisp, not fuzzy.

Assume that an automobile with activated cruise control has been traveling on a level road at a constant speed of forty-five miles per hour. The car comes to a hill and slows down. The cruise control comes to the rescue by automatically restoring the speed of forty-five miles per hour (the driver can intercede if he or she wishes).

The essence of automobile cruise control begins with Figure 5-1, which shows a fuzzy network with two crisp inputs and the resulting final output. The network is what will perform the magic. The automobile provides the following information concerning the input variables, as follows:

1. The vehicle speed, in rads per second (or radians per second), as measured by a tachometer (which senses the rotational speed of a shaft), and
2. The load-torque (a turning force) based on the percentage slope of the road.

The output of the network is the automobile throttle movement, whose job is to keep the automobile moving at a constant speed, in rotational units per second.

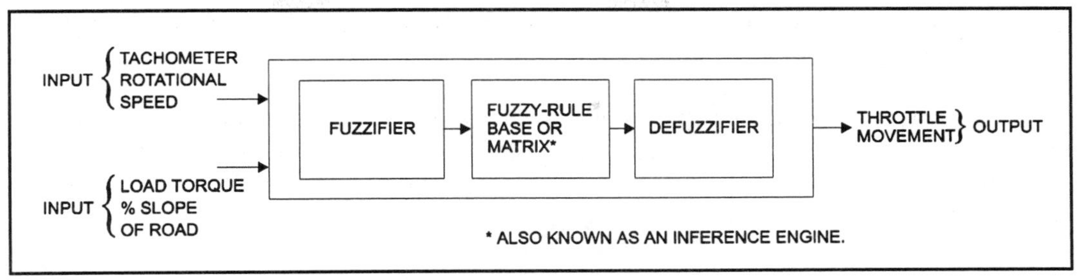

Figure 5-1. Fuzzy Network for Automobile Cruise Control

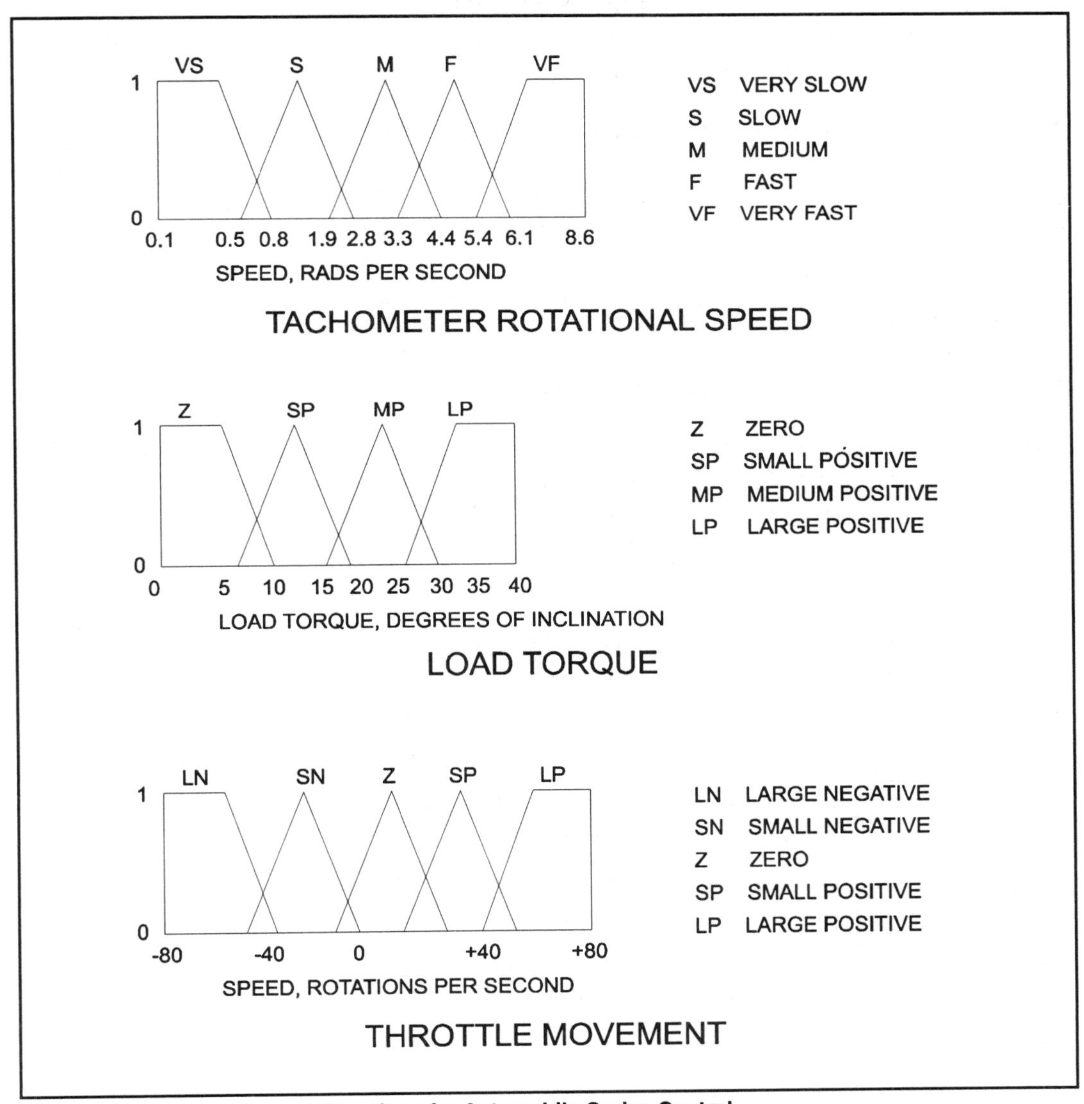

Figure 5-2. Fuzzy-Element Functions for Automobile Cruise Control

LOAD TORQUE \ TACHOMETER	VERY SLOW	SLOW	MEDIUM	FAST	VERY FAST
ZERO	LARGE POSITIVE	SMALL POSITIVE	ZERO	SMALL NEGATIVE	LARGE NEGATIVE
SMALL POSITIVE	LARGE POSITIVE	SMALL POSITIVE	ZERO	ZERO	SMALL NEGATIVE
MEDIUM POSITIVE	LARGE POSITIVE	SMALL POSITIVE	SMALL POSITIVE	SMALL POSITIVE	ZERO
LARGE POSITIVE	LARGE POSITIVE	LARGE POSITIVE	LARGE POSITIVE	SMALL POSITIVE	SMALL POSITIVE

Figure 5-3. Fuzzy-Rule Base or Matrix for Automobile Cruise Control

Each set of the two inputs requires individual *fuzzification*, which is intended to cover the entire range of automobile speeds and loads that may be encountered. Fuzzification turns the real-world input variables into linguistic variables that, in this case, will be the terms *Very Slow* through *Very Fast*, and *Zero* through *Large Positive* (see the text discussions referred to in Figure 5-2 and used in Figure 5-3).

Within each set are its two elements, which have a paired *membership value* or *degree of membership*. The membership is attached to a fuzzy-rule base, and all the other memberships are treated similarly for the remaining bases. Each set has a firing weight that is adjusted for firing as in Figures 5-2 and 5-3, and each has a numerical value, a degree of truth, within the range of 0 and 1 according to its importance.

Figure 5-2 shows that the tachometer reading of 0.2 rad per second fits into the *Very Slow* (VS) zone while the 2% load-torque fits into the *Zero* (Z) zone. The fuzzy-rule base determines which particular fuzzy-rule (see the text of Figure 5-3) is to fire, if firing is needed. To fire is to command an appropriate rule from the rule base to speed up or slow the throttle movement as necessary to maintain the proper automobile speed. (*Fire* is a verb used to describe a biological brain that may fire an electrical impulse; this was carried over to the artificial fuzzy brain.)

The entire Figure 5-3 is a fuzzy-rule base or *matrix*. Each of the twenty boxes inside the borders of the figure is designated as a *fuzzy rule*. A *defuzzification* develops a weighted average of all the fuzzy rules that contribute to the output. The *defuzzifier* then takes the fuzzy results from the fuzzy-rule base and turns them into an unfuzzy, real-world, crisp output, which is the automobile throttle movement. Whenever a fuzzy rule settles to *Zero*, the automobile is at a correct speed specified by the cruise control.

Examples of our automatic cruise control include a *Very Fast* tachometer with a *Medium-Positive* load-torque, which requires no corrective action, *Zero*, and a *Fast* tachometer with a *Large-Positive* torque, which requires a *Small-Positive* correction.

5-1-4 Benefits of Fuzzy Logic

Unlike crisp logic, fuzzy logic permits immeasurably greater flexibility and subtlety of performance compared to the limitations, the rigidity, of binary control. It permits small and large errors to be given different weights as needed to minimize over- and undershooting. Large errors may be given large corrections, and small errors may be given small corrections.

When a process is successfully controlled by a plant operator, an evaluation of his or her overall performance may lead to the process being controlled by fuzzy logic. By seeing what has worked, this may lead to improvements of the process through the consistent application of the best operating practice, and it could also reduce the need for operator intervention.

5-1-5 Successful Applications of Fuzzy Logic

The use of fuzzy logic is growing very rapidly worldwide. Some of the areas in which it has been applied are washing machines, electric razors, elevators, microwave ovens, a municipal subway, cardiac pacemakers, burners on a rotating kiln for manufacturing cement, the manufacturing of semiconductor chips, currency exchange rates, supervisory control, decision-making models, and scheduling.

5-2 How Neural-Networks Control

5-2-1 Neural Networks

The human brain is a wondrous tool. The concept of control by *neural networks* or, strictly speaking, by artificial neural networks (ANNs), which are also called *parallel distributing processing*, is based on mimicking some of the power and flexibility of the human brain. Research was performed on neural networks in the 1960s, and the networks became process control tools in the early 1990s. The purpose of neural networks is to solve certain control problems more efficiently than traditional methods can or problems that, in some cases, have been more or less intractable.

A natural, that is, a biological, *neuron* is a nerve cell that includes one or more input dendrites, output synapses, and more. The human brain has approximately one hundred billion neurons (100 E+09); each neuron may interact directly with 10,000 other neutrons, yielding a total of (metric) petaneurons (1 plus fifteen zeros). An artificial neural network is a dense mesh of nodes and connections that are primitively analogous to their counterparts in the human brain. A *node* is a point in a tree structure where two or more branches meet. By comparison with the colossal capabilities of the human brain, the artificial brain is pathetically puny.

Yet the artificial neuron can perform very useful work for us. It has some functional similarity to the natural neuron in that it can perform three basic functions: evaluate inputs and their strength, calculate a total of all the inputs and compare the total with a threshold level, and determine the proper output.

Neural networks can create process models that avoid the need for difficult calculations. They do not require any understanding of a process or its behavior. They "learn" about the behavior of a process by being given data that describe the input/output relationships in a process. The networks can also match test data closely. And they can perform computations efficiently and replace models that are less efficient, thus helping to make real-world applications practicable.

Models can be effective when used in the range in which they were fitted to real process data. Otherwise, they may not be reliable.

5-2-2 A Sample Network

There are various designs for neural networks. One common design is that of a feedforward neural network. Figure 5-4 shows a typical network that could operate as follows:

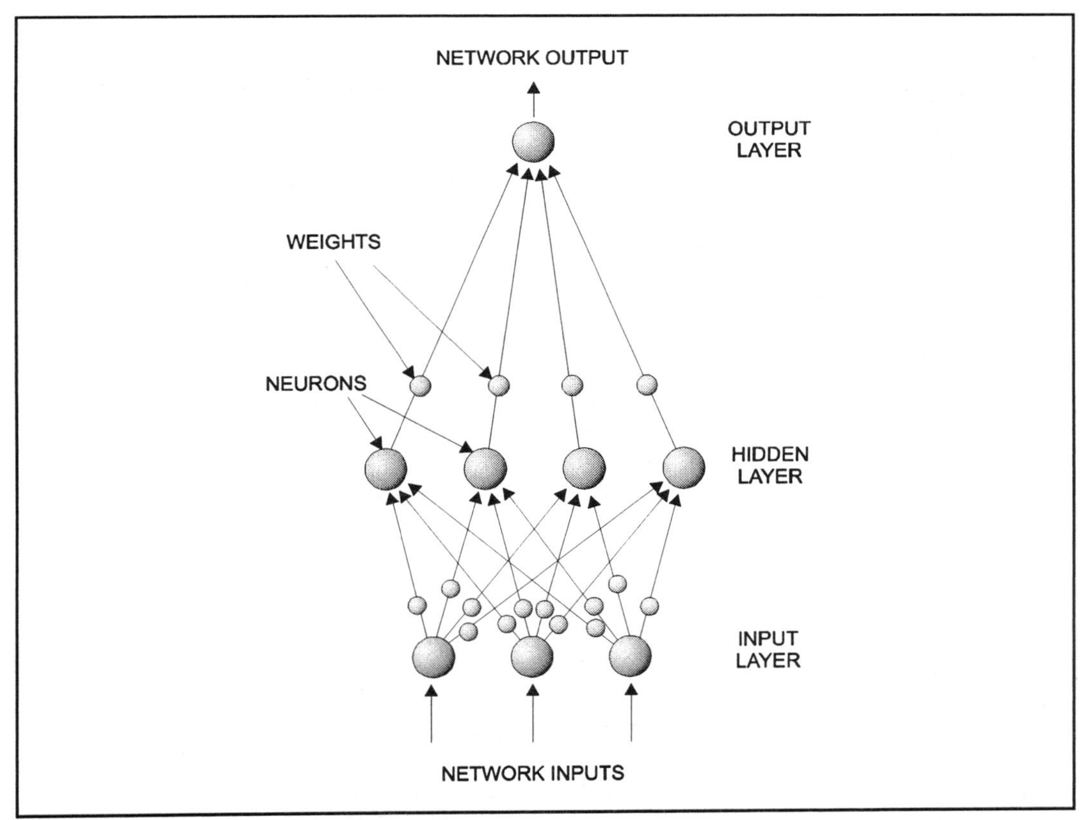

Figure 5-4. Feedforward Neural Network

- There are a number of network inputs from a process. The information from each input is distributed simultaneously to a group of neurons.

- The output of each neuron has an adjustable weight that is a measure of the relative importance of the various signals among neurons. All the weights must be adjusted properly to provide a true model. The weight adjustment requires an optimization search by one of various techniques. This adjusting is called *training* the network. When the network is trained, it has learned the process.

- The training must now be validated. A trained network has to be tested by a full range of data that it has not seen before. If the network has indeed learned to model the process behavior, then it will closely match the test data.

- To prevent overtraining, the training may be stopped just before the network begins to match the inevitable sampling noise. However, it may be possible to derive fundamental input/output relationships even when dealing with data corrupted by sampling noise.

- The output of all the weights goes to a single output, which is the network output.

- The outcome of a neural network is a prediction that is estimated from *past behavior*. After the network receives data that describe the input/output relationship in a process, the network can subsequently describe the behavior of the process.

- A trained neural-network model makes almost instantaneous predictions. The weight of a neuron changes by experience as the network learns from various inputs and follows its rules for modification. Predictions are unreliable if they are based on data that are outside the range of the training data or are collected from a modified process. In real situations, variables that are presumed to be stable over a short time period may drift over a longer period; this may necessitate periodic checking.

Neural networks may at times do a better job than other methods in relating a new situation to the most nearly similar past situation. This is because the network examines inputs concurrently rather than by a chain of logic statements.

5-2-3 Virtual Sensors

Many important process measurements, especially those involving the quantitative and/or qualitative analysis of substances, are difficult to obtain on line. On-line analyzers may be troublesome to operate and difficult to maintain, particularly in an awkward or dirty industrial environment, and they may be expensive. Yet reliable answers are required: what to do?

There may be an out. *Virtual sensors* are functional substitutes for conventional instruments that are difficult to apply. The virtual sensors are made up of conventional, simple, reliable, readily available, and moderate-cost instruments. Today, they are often used successfully in neural networks in the process industries. Many processes that require analyses are well instrumented or can be instrumented by a variety of conventional sensors.

Typically, such processes have an existing control system that has collected operating data for years. These data or newly developed data may be used to estimate the behavior of a difficult-to-measure or slow-to-respond variable on the basis of that variable's past pattern of behavior.

Such sensors may either supplant or complement analyzers or other problem variables. An example of this is *viscosity measurement*. Many plastics are manufactured in continuously stirred polymerization reactors, and the contents are quite viscous. The fluid viscosity (the degree to which a fluid resists flow) of the reactor contents is a major factor in controlling the quality of the final product. Control of the polymerization is based on the addition or reduction of chemicals when and as needed. Typical on-line viscometers take up to three minutes to provide an accurate reading, which is too long to make the required and proper additions of chemicals in time.

Viscosity can be inferred by the torque placed upon the electric motor that powers the stirrers inside the reactor. But the greater the viscosity, the greater is the torque required to stir the reactor contents, and this causes the electric motor to draw more electricity. Unfortunately, the motor-current signals are inherently very noisy and erratic, and they do not give reliable information.

A plastics manufacturer has been using neural control with one sensor to measure a motor torque, despite the signal noise, and a second sensor to measure viscosity by means of a viscometer that suffers from the lengthy time delay just described. The control system was trained by using recorded data from the plant production line. Once trained, the neural network predicted viscosity accurately despite the motor-signal noise and the viscometer time delay; immediate readings of the viscosity inside the reactor are now available. The result? The predicted viscosity is being used successfully by a process computer to control the flow of chemicals to the reactor vessel and to maintain the polymer quality.

5-2-4 Successful Applications of Neural Networks

Neural networks have been successfully employed for quality control, process control, credit card fraud detection, financial portfolio management, handwriting recognition, medical data classification, pharmaceutical development, the design of distillation columns, electric utilities, food processing, pulp-and-paper operations, and more.

5-3 How Generic Algorithms Control

5-3-1 Generic-Algorithm Control

Generic algorithms are methods used to solve mathematically difficult or seemingly intractable problems by a process of evolution. Such methods would be nominal counterparts of the natural evolution of living creatures. An algorithm is a mathematical model for solving a problem in a potentially indefinite number of steps. Algorithms use heuristically guided (trial-and-error) and random searches to develop solutions to problems.

A starting point for solving a problem requires a statement of problem constraints and a varied number of possible solutions. A fitness measure is derived for each trial solution to select the most promising candidates, and a percentage of the best solutions are retained to search further. Thus, new trial solutions are created. The process is repeated until an acceptable solution is reached.

Generic algorithms need not stand alone in handling problems. They may work together with fuzzy logic, neural networks, and classical methods such as binary logic and proportional-plus-integral-plus-derivative control, as appropriate, for developing solutions to problems.

5-3-2 Successful Applications of Generic Algorithms

Generic algorithms have been used to improve or resolve problems such as industrial optimization, multivariable modeling, navigation in uncertain environments, job shop scheduling, and others.

5-4 Combined Systems

Fuzzy logic and neural networks, though quite different from each other, are complementary. They may work individually or jointly to make use of each other's capabilities. This same intermingling may also apply to traditional means such as proportional-plus-integral-plus-derivative control, and binary logic. Whatever works may be used. This applies to both the hardware and the software of the system.

All added improvements in methods increase the overall capabilities and flexibilities of the process system if and when they are needed, and they also provide the control engineer with greater scope during design planning.

6

How Final Control Elements Function

Every control loop of whatever kind has to manipulate a flow of material or energy (see Section 1-2 and Figures 4-1 and 4-2). Though there are different ways to manipulate the flow, the predominant method by far is by means of a control valve.

6-1 Control Valves

We are all familiar with valves from our daily use of kitchen and bathroom faucets. These are simple hand-actuated devices for turning the flow on or off, limiting the flow, or switching the flow between the bathtub faucet and the shower head.

A *control valve* is a device, other than a common hand-actuated on-off valve, that adjusts one or more passages to change the flow rate or path, or both, for one or more fluid streams. In process control, there are *on-off valves, throttling valves,* and *modulating valves*. These are described as follows:

- An on-off valve, which operates in binary fashion, fully opens or closes one or more flow paths. It is an *isolation valve* or *shutoff valve* that may be used actively for manipulation (see Section 4-2-1).
- A *throttling valve* is used to limit flow by adjusting the opening. The limiting of flow is called *throttling* because the flow is choked. There are two different kinds of service, nonmodulating and modulating, which use throttle valves in two ways.

  1. *Nonmodulating Throttle Valve*. The valve is adjustable to different positions that remain fixed until the next adjustment; an example is a kitchen faucet. The adjustment is performed by hand, either directly or by remote control, or by a binary controller. When the valve is operated directly by hand, it is called a *hand control valve* (HCV); when operated remotely by hand, it is a *hand valve* (HV).
  2. *Modulating Throttle Valve*. This is usually called simply a modulating valve. The valve is operated by an automatic modulating controller that causes the valve to modulate in turn.

The subject of control valves can be surprisingly sophisticated and subtle. As a crude measure of this, the cost of a single valve for use in a process plant may range from five dollars to half a million dollars (for a nuclear power plant). Size differences are only part of the reason for the cost ratio being one to 100,000. There are other factors, which relate to performance, construction, materials, and quality.

A control valve is an assembly of hardware composed of three basic subassemblies:

- A body that is a pressure housing through which fluid passes. The body has connections for pipe or tubing.
- A movable *plug* that fits inside the body. The plug position changes to manipulate the flow. The plug may be a disk, a ball, or of some other specialized design.
- An actuator that is connected to the plug. It can move the plug when the actuator receives motive power.

These basic elements and the principles of valve operation are generalized in Figure 6-1. There are numerous subparts, including bonnet, packing, seat, stem, and others. The major considerations in choosing a control valve are discussed in the next section.

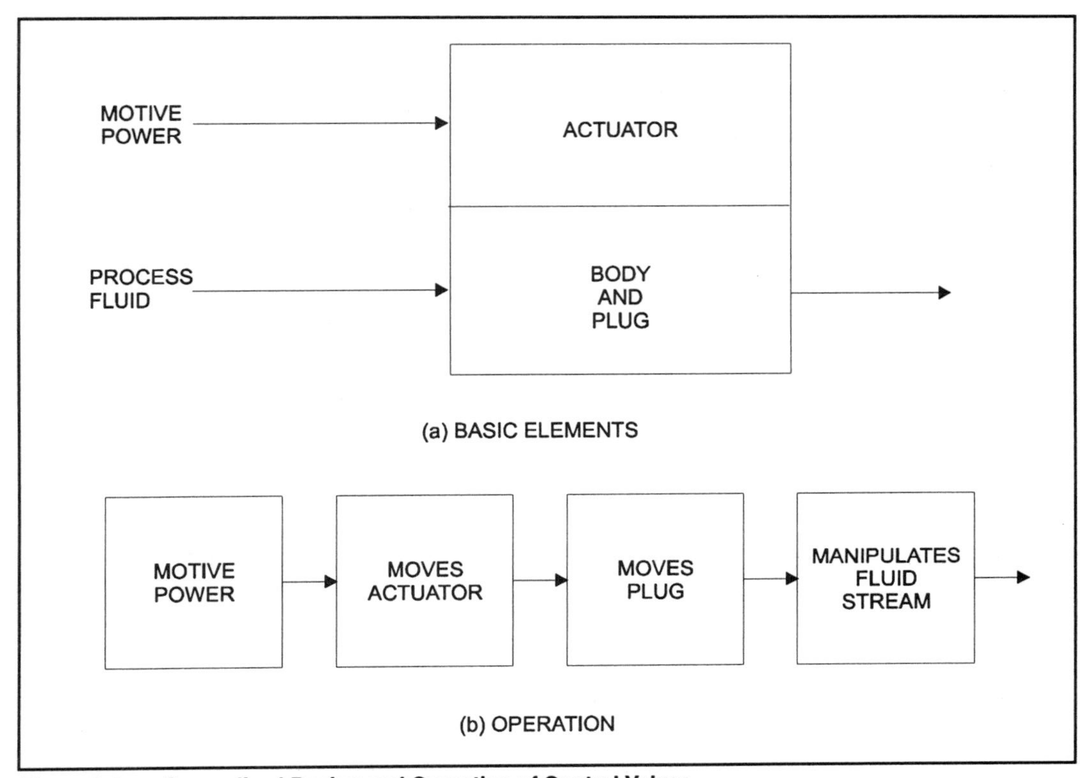

Figure 6-1. Generalized Design and Operation of Control Valves

6-1-1 Body Type

Examples of different *types, patterns,* or *styles* of valve bodies are as follows:

- *Globe.* This type has a plug that, typically, moves up and down to partially or fully open or close one or two openings, known as ports, as shown in Figure 6-2(a). The three-way valve in the figure is designed with two outlet connections; when the plug increases the flow to one of the outlets, it decreases the flow to the other outlet, and vice versa.

- *Angle.* This is basically like a single-port globe-type valve except for the orientation of one of the pipe connections. It is illustrated in Figure 6-2(b).

- *Butterfly.* This has a disk or vane that is mounted on a shaft that can rotate a quarter turn, ninety degrees. The disk is like a furnace damper except that it is much more sturdy and can be built for heavy-duty service. Figure 6-2(c) depicts one butterfly design, known as a *wafer* type, which has an abbreviated body for installation between a pair of pipe flanges. An alternative design has a longer body with a pipe connection on each side.

- *Ball.* In this type, passages are cut into a ball that is retained inside the valve body. The ball is on a quarter-turn shaft. Figure 6-2(d) shows typical flow arrangements of this type. See Figure 6-2 for valve styles.

6-1-2 Flow Characteristic

Every valve that throttles flow has a *valve characteristic,* which is the relationship between the amount of *valve position, travel, stroke,* or *lift* in the opening direction as compared to the resulting flow. The characteristic that is best for each application depends, as will be seen, on the specific control loop and the design of the entire piping system that contains the particular valve. There are three idealized types of characteristic, as follows:

- *Equal-percentage* type, for which equal increments of stroke to open the valve increase the flow by a constant percentage of the flow existing at the time of the change. The flow is not proportional to the stroke.

- *Linear* type, for which equal increments of stroke increase the flow by a constant amount. The flow is proportional to the stroke.

- *Quick-opening* type for which equal increments of stroke give decreasing increments of flow. The flow is not proportional to the stroke.

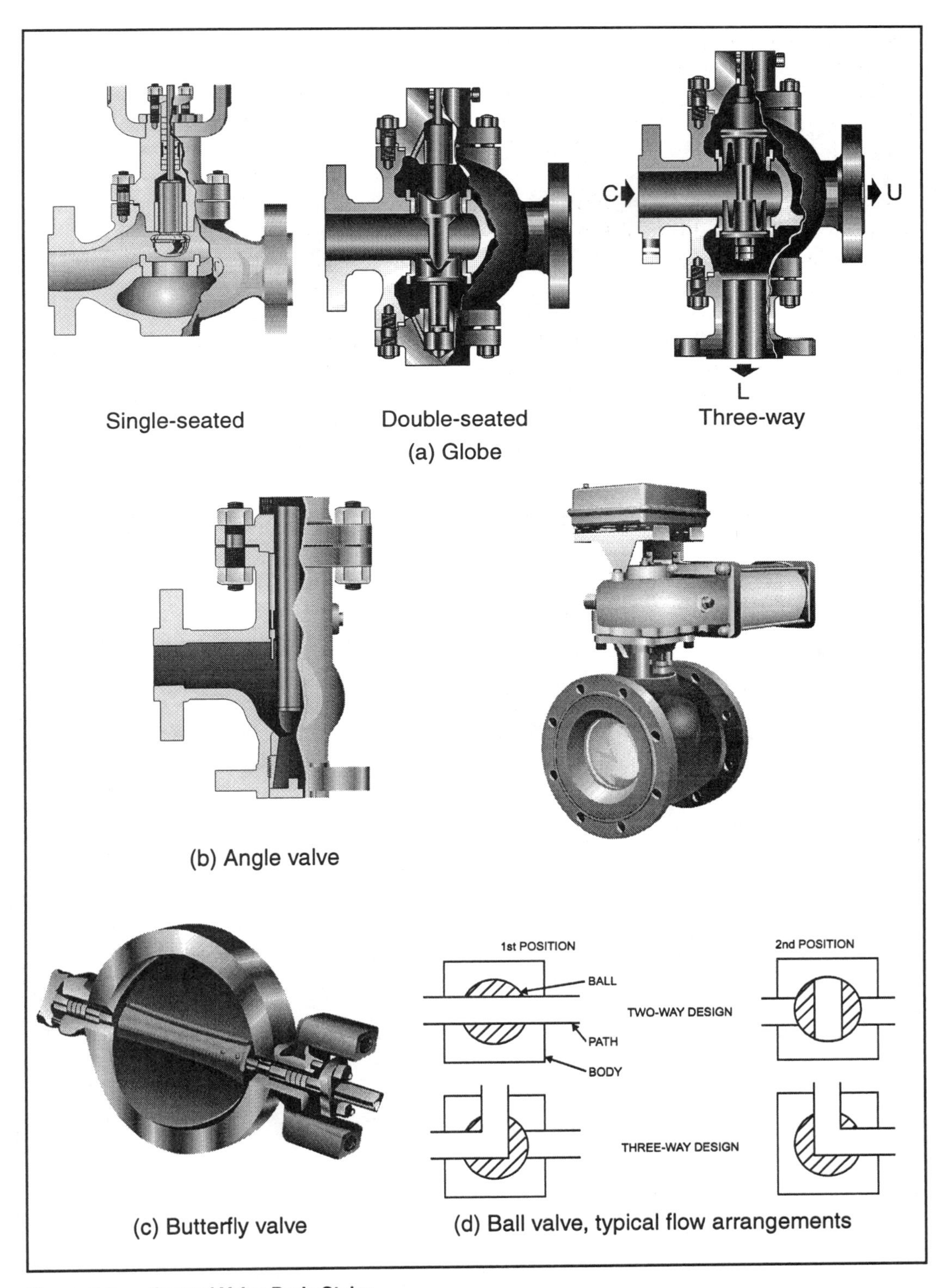

Figure 6-2. Control Valve Body Styles

The equal-percentage and linear types require specially shaped valve internals that follow mathematical equations for the desired relationship between stroke and flow. The equations assume that the *pressure drop*, which is the pressure difference between valve inlet and outlet, is kept constant so that the valve flow depends on only the valve stroke and characteristic.

The quick-opening type generally consists of a flat disk in a globe or angle valve, as in the traditional kitchen faucet. Some manufacturers modify the ideal characteristic. The characteristic as manufactured for all the types is called the *inherent characteristic*.

Valve characteristics are illustrated in Figure 6-3, which shows that an idealized control valve that has traveled 75% of its total stroke has a flow of approximately 42%, 77%, or 98% of its rated capacity, depending on whether the characteristic is equal—percentage, linear, or quick-opening, respectively.

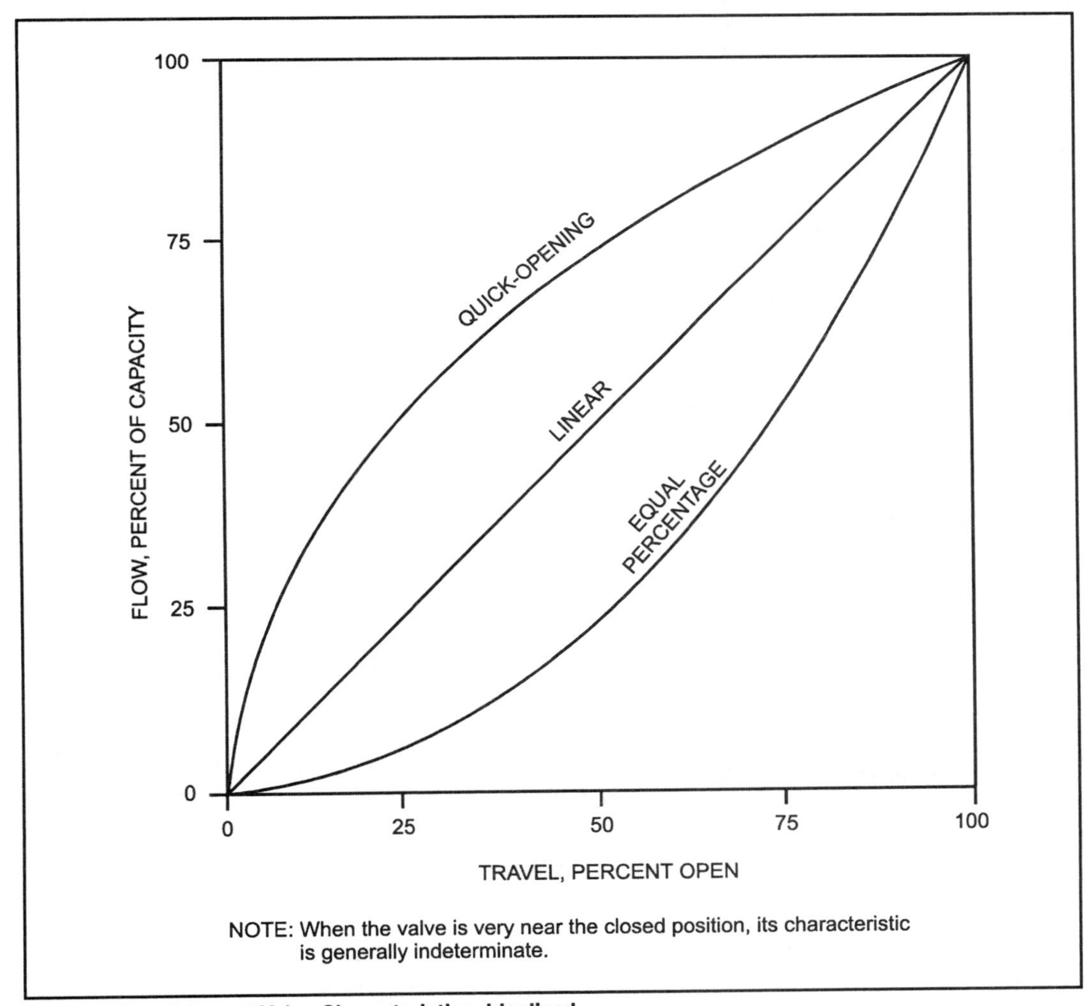

Figure 6-3. Inherent Valve Characteristics, Idealized

From Figure 6-3, we see that if the valve has an equal-percentage characteristic and we open the valve from 25% to 50% of its full travel, the flow increases from approximately 6% to 17% of capacity, a change of 11% of capacity. For a linear characteristic and the same valve movement, the flow increase is from 25% to 50% of capacity, a change of 25%. Then, moving the valve from 75% open to 100% open causes flow changes of approximately 58% for equal-percentage and 25% for linear. This is summarized in Table 6-1. Incidentally, the reason why most modulating valves do not control down to zero, as illustrated in Figure 6-3, is that there generally is uncontrolled leakage flow even when the valve is nominally in the closed position.

Table 6-1. Valve Characteristics: Equal Percentage Versus Linear

	Characteristic	
	Equal Percentage	**Linear**
For valve travel from 25% to 50% open, the flow change* is	6% to 17% = 11%	25% to 50% = 25%
For valve travel from 75% to 100% open, the flow change* is	42% to 100% = 58%	75% to 100% = 25%

*All percentages are relative to full scale.

We could go through a similar process to compare a linear characteristic with a quick-opening characteristic. However, from an inspection of Figure 6-3 and Table 6-1, we can draw conclusions, as shown in Table 6-2.

Table 6-2. Summary of Valve Characteristics

	Valve Characteristics		
	Equal-Percentage	**Linear**	**Quick-Opening**
Relative valve gain (Sensitivity):			
At low part of range	Lowest	Intermediate	Highest
At high part of range	Highest	Intermediate	Lowest
Gain change as valve opens	Increases	Constant	Decreases

Note: Valve gain equals the change in flow divided by the change in valve travel.

What is the point of having a choice of valve characteristics? After all, Figure 6-3 shows that each characteristic can handle flows from close to zero up to 100% of capacity. The answer is that the characteristics affect how controllers are tuned, and they can often make a significant difference in the performance of the control system.

An important factor for process controllability is this: Many pumps, heat exchangers, chemical processes, and control loops operate *nonlinearly*. This means that they require varying amounts of the manipulated variable to hold the controlled variable steady as the process load changes by equal amounts. The nonlinearities exist because the process capacitances and resistances change as the load changes. The nonlinearities are different for different processes. Figure 6-4 shows how the manipulated variable varies uniformly with load for a linear process and how it may vary nonuniformly for nonlinear processes. Many control systems are more or less tolerant of nonlinearity and are not much of a problem in that respect, but cases occur where the nonlinearity is troublesome. Selecting the proper valve characteristic enables the system to be more linear, therefore easier to control.

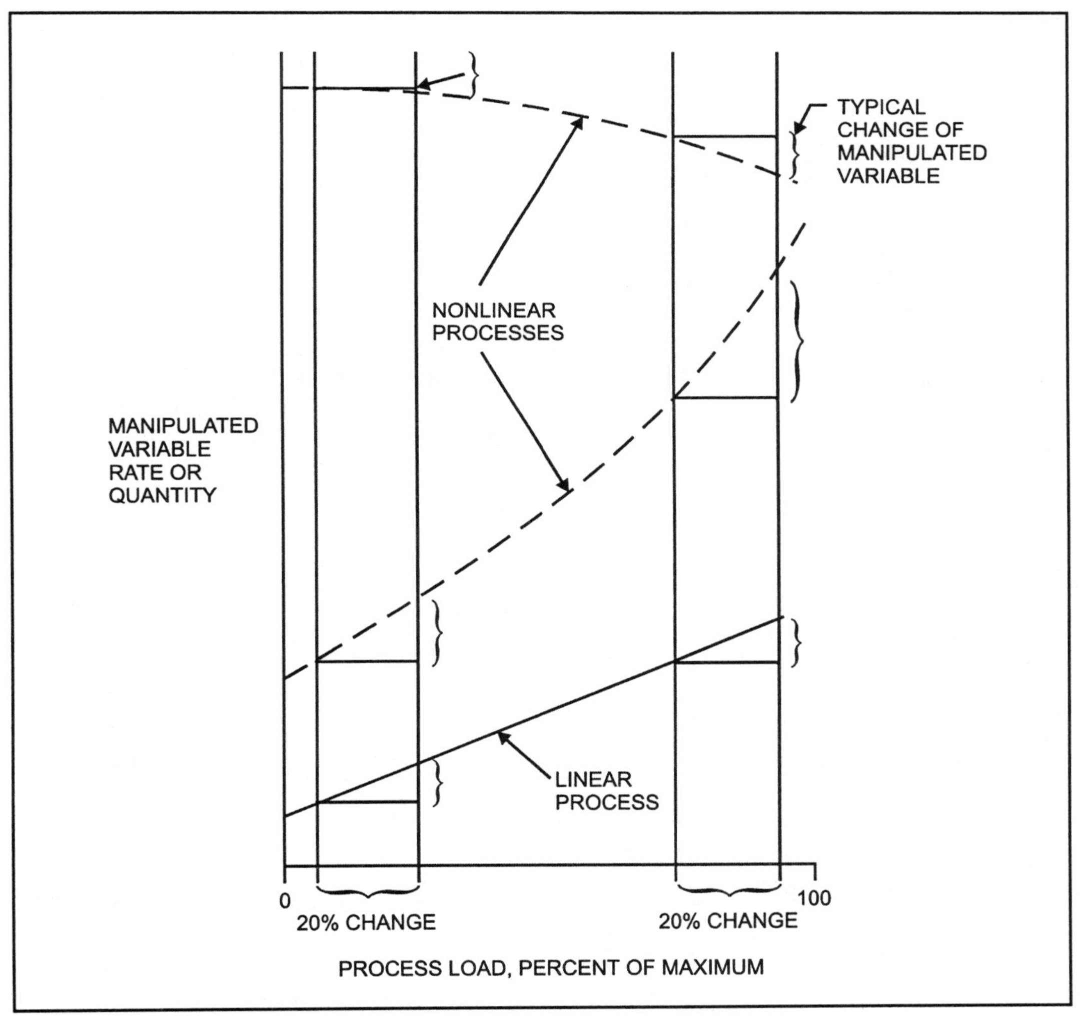

Figure 6-4. Process Linearities and Control

Tuning was described in Section 4-2-6 as the adjusting of a controller to provide the best possible performance for the control system. If the tuned system, which consists of the process and the control loop, is nonlinear, it is not possible to achieve the best performance for all process loads. If the controller is tuned at high load, it will be undersensitive and sluggish at low load. If it is tuned at low load, it will be oversensitive and unstable at high load. Tuning for a range of loads is easiest if the system is linear.

Another controllability factor relates to frictional pressure losses caused by flow in the piping circuit that contains the valve. Circuit pressure losses warp the valve manufacturer's inherent-characteristic curves shown in Figure 6-3 so that the equal-percentage curve bends toward the linear curve, the linear curve bends toward the quick-opening curve, and the quick-opening curve bends more to the left and becomes more extreme. These altered characteristics are called *installed characteristics.*

There are rules of thumb that simplify the selection of the characteristic for the majority of applications. The rules are based upon the following considerations:

- The kind of control that is involved—whether for flow, pressure, temperature, level, or other.
- The ratio of (a) the pressure drop of the valve alone to (b) the dynamic pressure loss of the entire piping circuit, all at maximum flow. Pressure drop for a valve is the difference between its inlet and outlet pressures. The valve drop for a given piping system cannot be assumed, and it varies with load. It equals the remaining pressure difference between the start and finish of the entire flow circuit after (a) deducting the total system *dynamic pressure loss*, which is the sum of all the friction losses for the entire flow circuit, and (b) correcting for the changes of pressure due to elevation changes, if any. The higher the ratio, the smaller is the difference between the inherent and the installed characteristics, and the easier is the control. The ratio varies with each installation.

The array of valve characteristics thus provides a choice for improving the process control by counteracting the problems of nonlinearity and piping-circuit pressure drop. However, other deficiencies in a given characteristic remain, as shown in the following list, so we do not expect to end up with a linear control system that is perfectly tuned over its full operating range. The other factors that affect the installed characteristic of a control valve include the following:

- Design approximations by the valve manufacturer.
- Manufacturing imperfections.
- Erosion of the internal parts of the valve, resulting in changes of the internal shapes.

- Later changes in design operating conditions.
- Boiling of liquid, known as *flashing*, which occurs inside the valves under some operating conditions.

There are accessory instruments, known as *characterizing relays*, that enable a nonlinear instrument signal to be linearized or otherwise shaped to improve system controllability or to be used for calculations. And there are control valve accessories that enable a given valve characteristic to perform as though it were different.

6-1-3 VALVE SIZE

For a valve, the word *size* needs clarification. *Body size* states the nominal diameter of the pipe connections, not the exact diameter. *Port size* refers to the diameter of the port. The port may be *full size*, meaning that it has the same nominal size as that of the body. Often the valve has a *reduced port*, also called *reduced trim* or *restricted trim*, meaning that the port is smaller than the body. *Valve size* may be stated as "2 in.", for example, which usually refers to the body size. Another valve may be "2 in. × 1 in.", which means that the valve has a two-inch body and one-inch trim.

Mathematical equations are used to size control valves, depending on the physical state and the operating conditions of the fluid. These equations express the flow capacity of a valve in terms of a valve-sizing coefficient, C_V. C_V is the number of U. S. gallons per minute of water at 60°F that can flow through the valve when it is wide open and the pressure drop is one psi. A larger C_V means a higher flow capacity. Though water is the reference fluid, C_V is used for valves handling all fluids. Manufacturers' catalogs list the C_V ratings for their various control valves according to the body style, body size, port size, and other factors. When a valve C_V is specified, it is understood that this is for a fully open valve, unless stated otherwise.

A rule of thumb for preliminary sizing states that the proper valve size is one size smaller than the pipeline size; for example, this suggests a three-inch valve in a four-inch line. However, the size must not be considered final until it is actually calculated from the appropriate equation. In a well-designed process system, the valve body size does not exceed the line size, and it is not unusual for it to be more than one size smaller.

The three-inch valve may be as adequate as a four-inch valve for manipulating the flow, and it would cost less. More importantly, the smaller valve very often controls better. A valve that has a greatly excessive capacity has operating disadvantages, especially for throttling service. The oversized valve, compared to a properly sized valve, throttles with the plug nearer the closed position, causing the following bad effects:

- *Excessive Gain.* A slight stroking of the oversized valve causes a relatively large valve opening change, with a correspondingly large change of flow. In other words, the valve gain, which is the ratio of (a) the change in flow to (b) a change in valve position, is high. High valve gain together with any erratic error in positioning the valve—and some error is expected, even with a positioner—magnify the resulting flow error; stable control is more difficult to achieve. This would be comparable to trying to have very fine control of the water flow in the kitchen sink if we replaced the faucet with a three-inch valve.

- *Effect on Flow Characteristic.* Because of design and manufacturing limitations, uncontrollable leakage flow generally exists and causes the planned characteristic of the valve to be indeterminate at very low valve positions. This effect becomes aggravated if the valve parts erode or corrode. Leakage flow may become important when the total flow is small. In general, the larger the port size, the greater is the leakage and the worse is the resulting control at low flow.

- *Dynamic Instability.* When a valve, such as a globe valve, is closed, its plug rests on that portion of the valve port known as its *seat*. In the near-closed position of the valve, the smallest area of fluid flow is the relatively small space between the plug and seat. This flow constriction causes a local increase of fluid velocity with a lowering of pressure, as shown in Figure 3-17 for an orifice plate. Especially for high-pressure drops, the lowering of pressure at the valve seat can have a strong sucking effect on the plug and cause it to slam against the seat. The sucking is the so-called *bathtub-stopper effect*, which occurs when a bathtub is being drained. If a rubber stopper is placed in the tub outlet by hand to stop the flow, the stopper is forcibly sucked down when it is very close to the outlet, and the stopper slams closed. For a control valve in modulating service, such stoppage of flow eliminates the downward pull on the valve plug, and the valve actuator reopens the valve to maintain the required flow. With the resumption of flow, the plug is promptly pulled down again while the plug is still close to the seat; this close/open cycling goes on and on. This action can be mechanically damaging to the valve and is upsetting to the process.

  One way to overcome this effect is to install a larger actuator to overcome the dynamic forces that cause the plug to cycle. This is a brute-force method that may increase the valve cost and may cause external space problems. It is better, where practical, to minimize the dynamic cycling by avoiding excessive oversizing of the valve.

- *Erosion.* There is a greater tendency for erosive wear of the valve because the plug is closer to the closed position, causing the local fluid velocity to be higher.

It is sometimes advisable to provide a larger body than is calculated on the basis of capacity. This may be because of the following considerations:

1. The fluid velocity entering or leaving the body may be excessive, thereby causing excessive erosion and noise. Enlarging the body reduces the inlet and outlet velocities.
2. If flashing and cavitation (boiling and recondensation) occur as a result of internal velocity changes, a larger body reduces the severe erosive effect of collapsing bubbles on the body wall. Also, the wall is thicker, with a longer life.
3. A structurally stronger body may be needed to withstand piping forces and vibration.
4. The costs of pipe reducers and their installation can be avoided by using a line-size body.
5. A larger body may be desired to allow for future plant expansion and a higher flow, but the port can nevertheless be sized to meet current needs by using reduced trim. A calculated three-inch valve may become a four-inch-by-three-inch valve with a body enlarged to four inches but with a proper three-inch port. Whenever required, four-inch trim can be installed in the four-inch body, and the piping will not have been disturbed.
6. The port may be oversized for current needs because of either an initial sizing error or a change in the valve operating conditions. The oversizing can be corrected by installing reduced trim.

Rangeability is an important concept for control valves. *Inherent valve rangeability* is usually defined as the ratio of the maximum flow that is controllable by a valve to its minimum controllable flow. The maximum controllable flow is the valve capacity, which is, of course, greater than the maximum flow requirement or else the valve will be a process bottleneck. The minimum controllable flow is not so clear-cut: it may depend on how large the leakage flow is—this gets worse with wear—or on how low the flow is where the valve characteristic becomes deviant. The inherent rangeability of types of valves can be stated only as approximations, with some common values as stated in Table 6-3. A valve in actual service has an *installed rangeability* that is smaller than the inherent rangeability because of pressure losses in the piping.

Table 6-3. **Nominal Inherent Rangeability of Typical Control Valves**

Characteristic	Stroke Resolution* 1%	10%	20%
Quick-opening	50:1	6:1	3:1
Linear	100:1	10:1	5:1
Equal-percentage	50:1	50:1	40:1

*Stroke resolution refers to the valve's positioning accuracy. It is of particular importance for defining the minimum controllable flow, which exists when the plug is close to the seat.

Process rangeability is the ratio of the maximum required C_V to the minimum required C_V, usually corresponding to the maximum and minimum flows, respectively, that are to be manipulated, with each flow being at its own operating conditions. When considering a control valve for a specific service, its installed rangeability should be larger than the process rangeability. Otherwise, the valve may be called on to manipulate flow over a range that is larger than it can handle. In such a case, a different style of valve may be called for, if otherwise suitable, or it may be necessary to use two control valves in parallel.

6-1-4 Actuator

The *actuator* is the drive or power element that positions the part of a final control element that directly manipulates the fluid flow or other manipulated variable. In the case of a control valve, the actuator positions the plug. (The word *actuator*, referring to the drive, is much preferred to *operator*, which should generally be reserved for a person. A person *operates* and an actuator *actuates*. However, it is also correct to say that a controller, for example, operates the actuator of a valve.) The parts of a typical actuation system are shown in Figure 6-5.

There are different kinds of actuators, based on the requirements described in the next three sections.

6-1-4-1 Type of Movement

A globe valve, for example, requires an actuator whose output movement is *rectilinear*, or simply *linear*, to move the valve plug up and down. By contrast, a butterfly valve requires an actuator that has a *rotary* output to rotate the valve disc. The valve movement begins with the drive, whose input movement is not necessarily like its output movement. The globe valve may be moved up and down either by manually rotating a knob that rotates a threaded valve stem, as is common in the home and industrially, or by pushing or pulling the valve stem, as is common industrially for control valves. The butterfly valve may be rotated by a rotating motor, usually electric, or by a push-and-pull actuator whose linear movement is converted into rotary movement by a mechanical linkage, as for a fireplace damper.

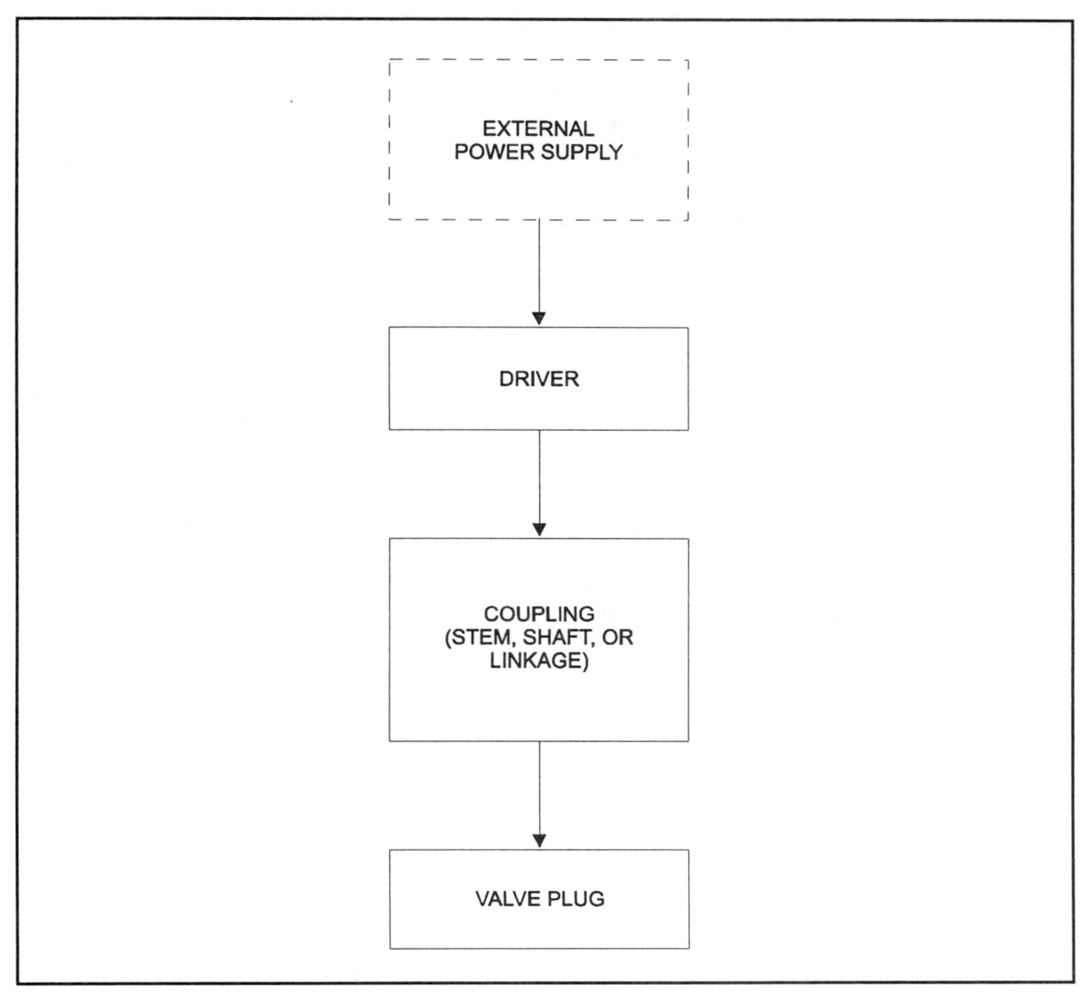

Figure 6-5. A Typical Control Valve Actuation System

6-1-4-2 Control Action

Actuators are used for on-off or modulating service. Electric actuators, such as motors and solenoids (which are basically electromagnets), may overheat and burn out depending on how frequently they are energized and how heavily and for how long they are loaded.

6-1-4-3 Power Output

To move a valve plug or other device requires that the actuator provide sufficient output force, which may be considerable for the mechanical parts of large valves. Additional actuator force is needed to overcome the forces of the process fluid. The fluid forces may be very large because of (a) a static pressure difference between the inlet and outlet of a vertical pipe, resulting in an unbalanced force, and (b) dynamic effects existing when there is flow. It may require more force to modulate a flow and keep it steady than to merely open the valve fully or close it. Another factor is how fast the valve is required to open or close. There are many processes for which speed of operation is unimportant; this may be true of a slow-changing process. In some cases, especially for emergency services involving safety, speed of operation in opening or closing may be very important. Of course, the greater the speed requirement, the more powerful and more expensive the actuator must be. Some turbine shutoff valves, for example, are twenty inches in size or larger; handle high-pressure, high-temperature steam; and are massively large and heavy. They full-stroke closed in 0.1 second; that takes a lot of power.

The force that an actuator can apply is described as follows:

1. For linear motion: The force may be stated as a "stem thrust of 1200 pounds," for example.
2. For rotary motion: This needs a twisting force, which may be stated as a "torque of 600 pound-feet," for example. *Torque* is the twisting force that makes a bolt turn when we use a wrench on it as a lever. The value of 600 pound-feet for the twisting force is the product of a linear force, in pounds, multiplied by the length of the lever, in feet. The SI unit for torque is the newton-meter, which equals 0.7375 pound-feet.

Rotary actuators may also have a power rating, such as "one horsepower," which depends on both torque and speed. When an actuator moves a valve through its stroke, the actuator force is exerted over the stroking distance. The force multiplied by the distance is the work done by the actuator. The work divided by the stroking time is the power capability or requirement for the actuator. The basic SI unit for power is the watt (W); one kilowatt (kW) equals 1.3410 horsepower (hp) or 1.3596 metric horsepower.

Similarly, a car that is driven up a hill at high speed requires more power than at low speed even though the work performed is the same in both cases. The work is performed by the car raising itself from the bottom of the hill to the top. Work does not depend on time; power does.

Altogether, the actuator must have the power and the force to provide the following positioning results:

1. Accuracy despite unbalanced fluid forces, the weight of parts, and frictional resistance.
2. Stability despite process pressure fluctuations.
3. Speed of operation as required for the application.

The following sections provide basic descriptions of common types of actuators.

6-1-4-4 Fluid-Powered Actuators

The great majority of control valve actuators in the process industries are fluid powered using compressed air. For on-off service, the most commonly used air supply pressures are approximately 50 psig, 80 psig, or 100 psig. For modulating service, common air supply pressures are 20 psig or 35 psig, and typical air-control-pressure ranges are 3-to-15 psig or 6-to-30 psig. In some cases, the power medium is hydraulic liquid at pressures of 3000 psig or higher. The fluid-powered category includes the three types of actuators described in the remainder of this section. Miniature versions are used for pilot valves that are sometimes used to control the operation of process control valves.

1. *Diaphragm Actuator.* This has a flexible diaphragm with a partial backing plate that is rigid. The air pushes one way, and an opposing spring pushes the other way. A valve stem is connected to the plate. The stem position varies with the air pressure. This is illustrated in Figures 6-6(a) and (b), which show control valves that move down and move up, respectively, as the air pressure increases. These are known as *fail-open* and *fail-closed* valves, respectively, because if control air to the actuator is lost the spring causes the (a) valve to open and the (b) valve to close.

   Both of the valves are *fail-safe,* meaning that they fail in a safe direction. For example, suppose a temperature controller is modulating the fuel gas flow to a furnace and the control air line to the control valve is damaged and breaks, causing the actuator air chamber pressure to fall to atmospheric pressure. The spring makes the valve fail to the safe position.

   But which is the safe position, open or closed? If the valve were to fail open, there would be an unrestricted flow of fuel gas to the furnace. This could be very dangerous. If the valve were to fail closed, the furnace would be starved of fuel, and the fire would go out; production would stop but there would be no danger. The choice is clear: We specify that the valve shall fail closed, as does the valve in Figure 6-6(b). Nobody is hurt, we repair the broken air line, and we are back in business.

   There are four possible modes of actuator failure: *fail open, fail closed, fail locked* (in the last position), and *fail indeterminate* (difficult to predict). ISA standard S5.1 has established standard symbols covering the four modes.

2. *Piston Actuator.* This uses a piston and cylinder. Figure 6-6(c) illustrates one version, called *single-acting,* that uses air pressure on one side of the piston and an opposing spring on the other side, similar to the arrangement for the diaphragm actuator, described previously.

   Figure 6-6(d) shows another common version, called *double-acting,* that uses no spring but has control air on both sides. This requires the use of a pilot valve or a positioner, which are accessories described later in this section. An input signal to the accessory causes the opposing output air pressures to vary in reverse to one another so as to balance the fluid forces on the control valve plug and push the piston up or down as required.

Regarding fail-safe action, the single-acting piston actuator functions like the diaphragm actuator. But the double-acting piston requires a small auxiliary subsystem that includes an air reservoir for emergency use. There are other variations.

3. *Motor.* This is a pneumatic rotary machine with gearing that can be used for on-off or modulating actuation of valves and other devices.

6-1-4-5 Electric-Powered Actuators

Electric-powered control valves are used for process work less often than are pneumatic-powered valves. Electric-powered types are as follows:

1. *Motor Actuator.* This is essentially an electric rotary motor with gearing that is used mainly for on-off control valves and other devices. Some are used for modulating service, especially in heating and air conditioning systems.
2. *Solenoid Actuator.* A solenoid is an electromagnet that has an annular (ring-shaped) electric coil, inside of which is a magnetic iron plunger that is connected to a valve stem. Applying voltage to the coil pulls the plunger into the coil. A spring returns the plunger to its initial position when the electric power is shut off. The stem movement is linear. Thus, the valve opens or closes, depending on the design, as the actuator is energized.

   Though solenoid valves can be designed to modulate by varying the power supply voltage, their major use is for on-off service. Solenoid actuators produce relatively little thrust and have limited use as control valves. They are used mainly as pilot valves for process control valves.

6-1-4-6 Manually-Powered Actuators

A manual actuator may be a handwheel, which has a rim, spokes, and hub. The wheel may be modified or replaced, as follows:

1. There may be no rim, and the number of spokes may be reduced to two or even one.
2. There may be no rim and no spokes. The hub has wrench flats for use with a wrench.
3. The rim may have a chain to permit manual operation from a lower elevation.
4. The actuator may include a set of gears to (a) reduce the force that the operator must exert or (b) change the type of output motion from rotary to linear.

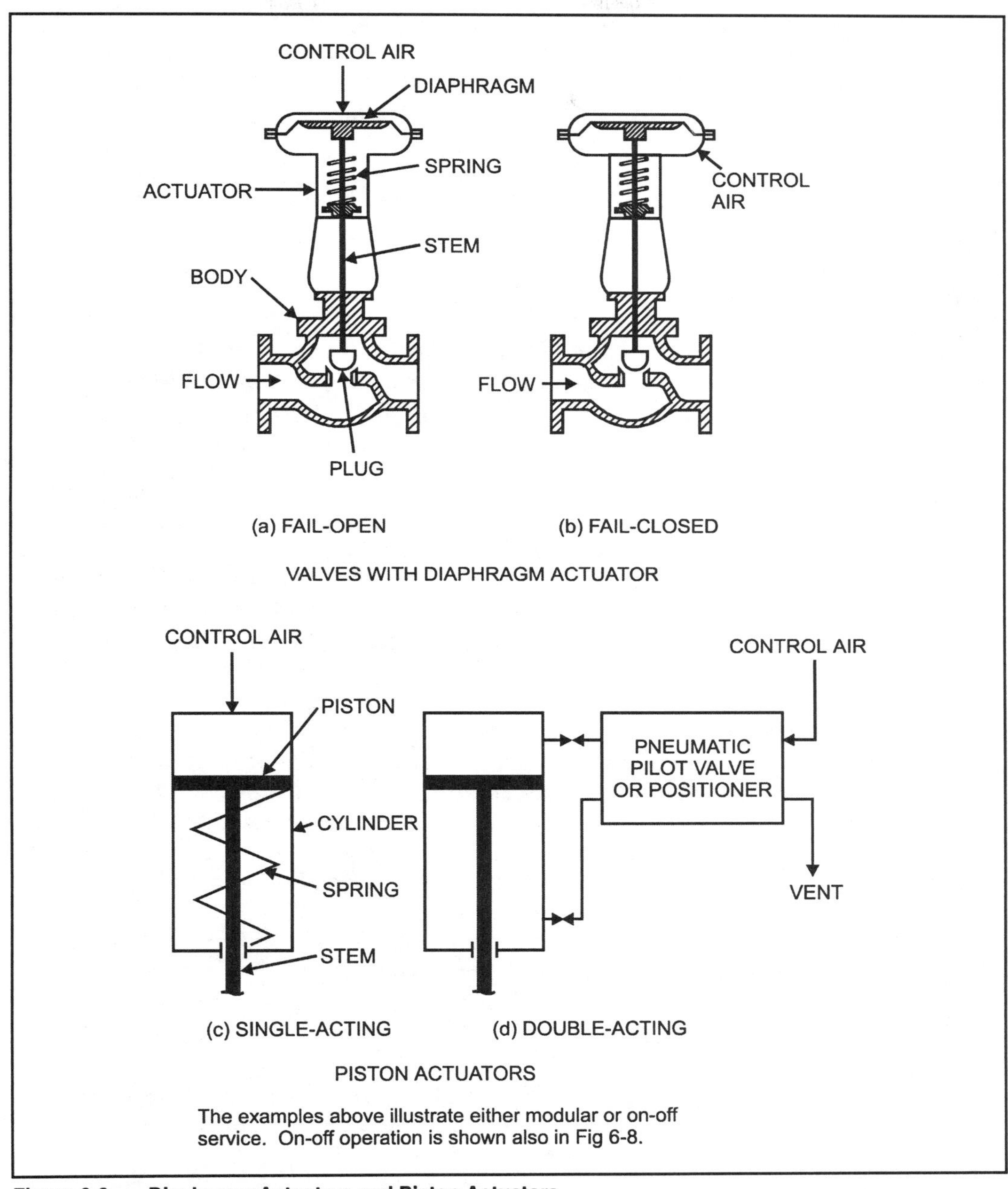

Figure 6-6. Diaphragm Actuators and Piston Actuators

Manual actuators alone are used only when the remote or automatic operation of a valve or other device is not required. Their major use is for the isolation of equipment or systems. They may also be used to throttle flow where the process is slow changing and their relative unreliability, because they depend on a human operator, is acceptable. (See Section 1-3-4, "Process Reliability and Safety.")

A manual actuator may be attached to an automatic valve to override the automatic actuator. This may be done to permit continued plant operation if the automatic control system is out of order. This manual actuator has a clutch for disengaging the automatic actuator, after which the operator has full manual control of the valve.

A manual actuator, known as a *jack* or *hand jack*, may be used not to override the automatic actuator but rather to limit the range of travel of the valve. It is an adjustable travel stop. The jack is threaded and may have a handwheel or wrench flats.

A valve stem may be lengthened by placing a stem extension between the manual actuator and the valve body. This permits the manual operation of the valve from a distance.

6-1-4-7 Positioner

A major accessory for many modulating control valves is a *positioner*. This is a position controller that senses valve position directly by means of an external mechanical connection to the valve stem. The set point of the positioner is valve position, as commanded by a process controller, such as a temperature controller. The positioner automatically adjusts the signal to the valve actuator if the actual position does not match the set point. The position controller is part of a cascade control loop (see Section 4-3-2), which is illustrated in Figure 6-7.

Figure 6-7 shows a system for cooling hot water by mixing it with cold water. The temperature of the tank effluent is controlled by a modulating temperature controller that controls a positioner that in turn operates a control valve that itself manipulates the flow of cold water. The temperature controller is the primary controller of the cascade loop; the positioner is the secondary controller.

The valve could operate without the shown positioner, but, in this particular case, it would not operate as well. The tank has a large thermal capacitance; therefore, temperature changes are slow. Because of friction and unbalanced fluid forces in the valve, a control valve without the positioner would not take *exactly* the position that corresponds to the temperature controller output signal. There is a valve-position error that causes an improper flow of cold water. The position and flow will not be corrected for a relatively long time for the following reasons:

1. The process is slow.
2. The flow error is small.
3. No adequate correction will be made until the temperature error has built up again to change the temperature controller output sufficiently to overcome the forces that restrain the valve movement.

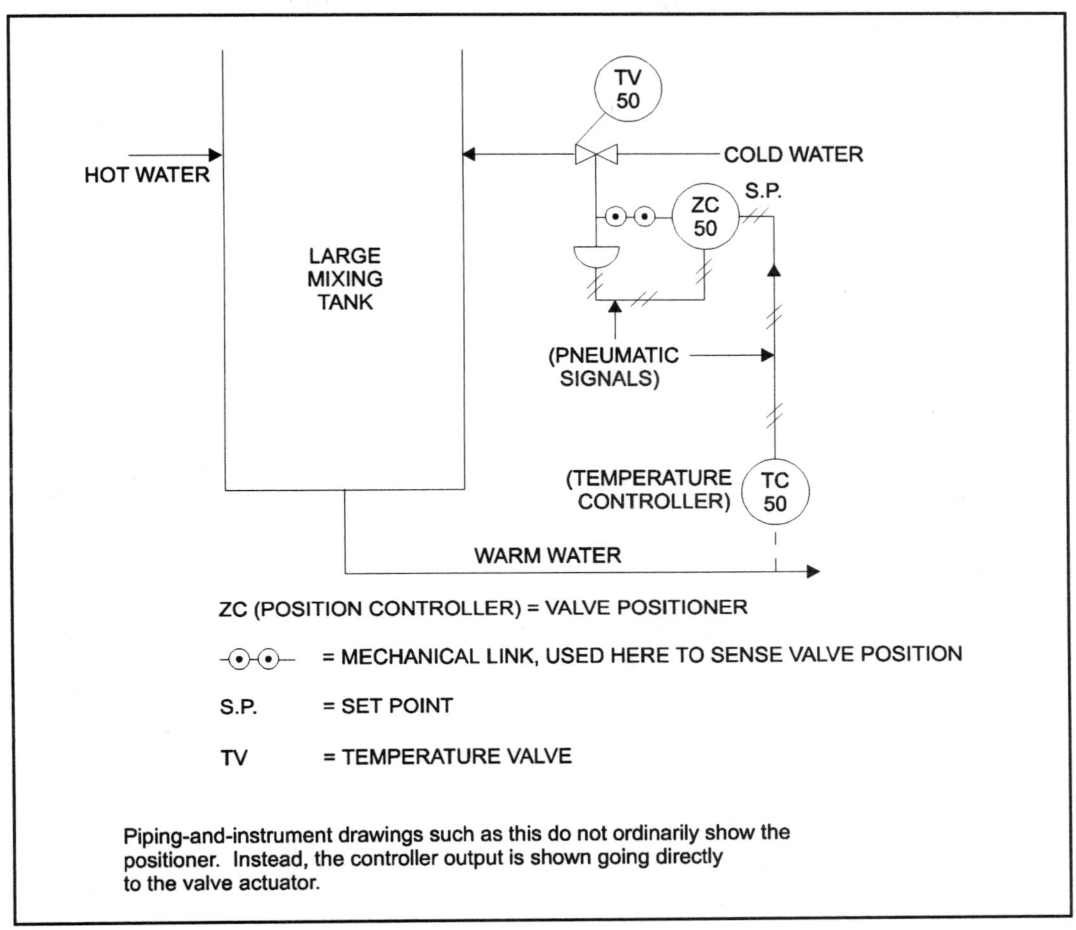

Figure 6-7. Using a Valve Positioner

Finally, the valve position is readjusted, again perhaps not exactly correctly. The point is that the control error had to keep growing for a while because the valve responded imperfectly to the command of the temperature controller. The control error will be greatly reduced by having a positioner on the valve.

The rule is that a positioner is not used unless it is needed. It is needed only for modulating control in the following cases:

1. If the process is slow, as is common for the control of temperature or for analysis, for example. Control is actually degraded if a positioner is used for a fast process, such as most pressure- or flow-control systems, regardless of how bad the valve friction and unbalanced forces may be. This follows the general rule for cascade systems, namely, that the secondary controller should be installed only if the primary loop is slower than the secondary loop (see Section 4-3-2). In this case, the positioner should be used because the control valve can change faster than the

process can. This practice differs from the traditional but obsolete criterion for using a positioner to overcome forces that impede accurate positioning.

2. If a double-acting actuator is used, regardless of the process speed.
3. In other special situations.

Valve travel may be speeded up for a modulating pneumatic valve that does not require a positioner by adding a booster relay at the valve. The signal sequence is from controller to booster relay to valve actuator. A *booster relay,* or *volume booster,* is an instrument whose output pressure equals its input pressure but whose output air-flow capacity is larger than that of the input. Thus, it speeds up the pressurizing or venting of the sizable air chamber of a valve actuator.

Double-acting piston actuators in on-off service do not use a positioner but rather a small four-way valve, as shown in Figure 6-8(a). The valve moves the actuator fully up or fully down by switching air pressure from one side of the piston to the other. This valve may be hand actuated but frequently is solenoid actuated or pneumatic actuated. The valve is named a *pilot valve* because it does no work other than to direct the action of the control valve actuator. Figure 6-8(b) shows a three-way pilot valve used for a single-acting piston actuator.

6-1-5 MATERIALS OF CONSTRUCTION

There is a large choice of available construction materials for valve bodies and their internals. The internals are known as *trim*. When specifying valve materials, the following factors must be considered: structural strength, corrosion, and erosion.

6-1-5-1 STRUCTURAL STRENGTH

The material of the body and piping connections is normally specified by the purchaser. The body and connections must be made strong enough to withstand the following two forces:

1. *Internal pressures* in combination with high or low temperatures. The greater the pressure, the stronger obviously must be the material. However, the strength of any metal (a) decreases as its temperature rises and (b) is limited by brittleness and potential fracturing at cryogenic (very low) temperatures. The American Society for Testing Materials (ASTM) has established standard specifications for metals, and the American Society of Mechanical Engineers (ASME) has established allowable pressure and temperature working limits for different uses and types of metals. For example, a specific grade of carbon steel (ordinary steel) has a lower limit of –20°F (–29°C) and an upper limit of 1000°F (538°C); a specific type of Inconel (a high-nickel steel alloy made by the International Nickel Company) has limits of –325°F (–198°C) and 1200°F (649°C). The requirements have been systematized into ANSI standard pressure-temperature ratings. For example, control valve bodies come in standard

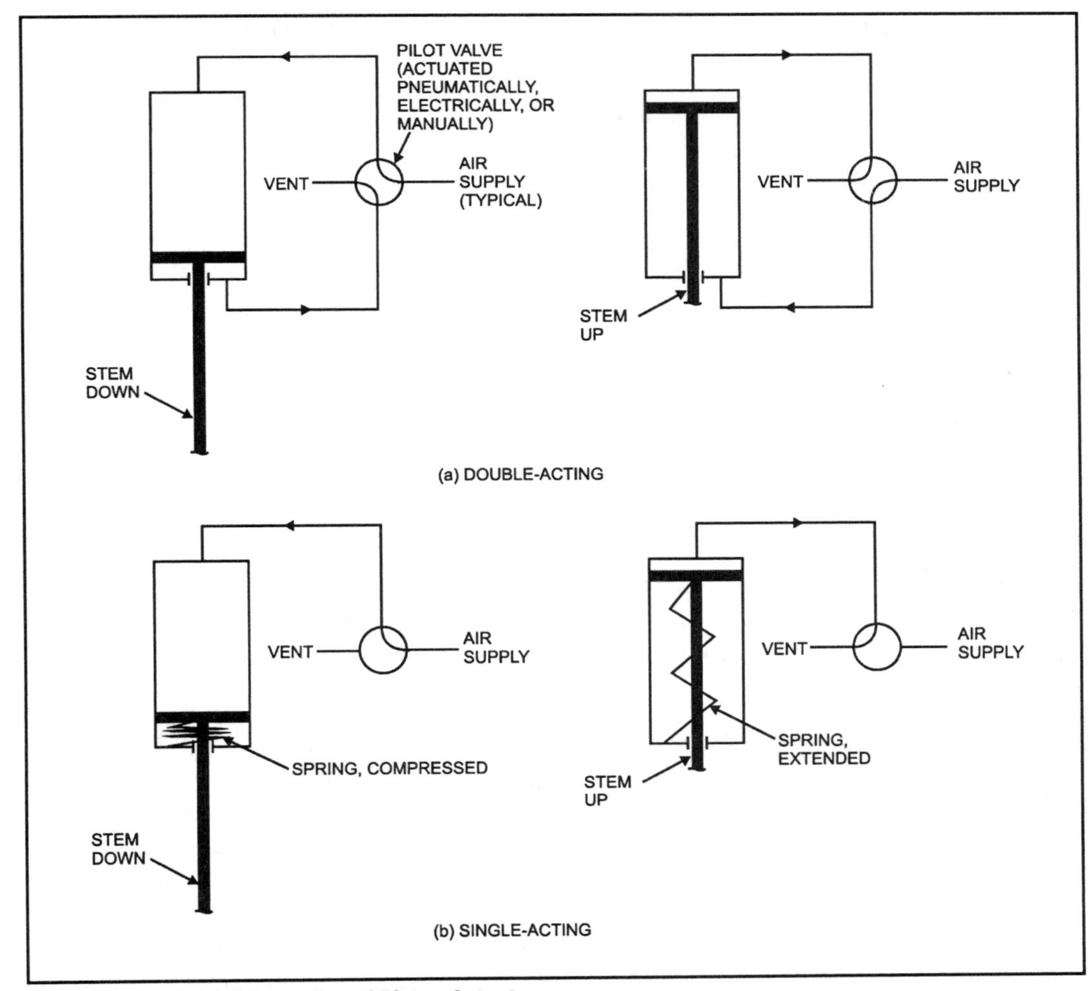

Figure 6-8. On-Off Operation of Piston Actuators

classes, 125, 150, 250, 300, 600, 900, 1500, 2500, and 4500. Each rating has a maximum allowable working pressure that is equal to the rating for a specific metal at a specific service temperature. At higher or lower temperatures, the allowable working pressures vary.

To specify the proper class rating for a valve, we determine the extreme combinations of pressure and temperature to which the valve may be subjected, and we select the lowest class rating that is safe for the worst combination. This rating will be the lowest in cost that is technically acceptable, except that we may choose a higher rating if corrosion or erosion of the body wall of the valve by the process fluid may later weaken the body. The standard pressure-temperature classes do not apply to nonmetallic bodies. For those, the valve manufacturer should be consulted.

2. *External forces* that the body is subjected to because of temperature changes in the piping or equipment to which the valve is connected. These forces may distort the valve body by compression, expansion, bending, or twisting. The result may be damage to the valve or interference with its operation by changing the clearances of internal parts. However, proper design of the piping by piping designers, including general adherence to good piping practices and the valve manufacturer's installation instructions, protects against the external forces.

Commonly used body materials are brass, bronze, cast iron, carbon steel, and stainless steel. There are also other, more exotic alloy materials.

The valve trim needs to have adequate strength to withstand fluid forces and the actuator force, but the valve manufacturer normally has the responsibility of providing the strength by using appropriate trim materials.

6-1-5-2 Corrosion

Process plants frequently use fluids that are very corrosive. Even water, as we know, is corrosive to iron and steel. If the wrong construction material is used for a control valve, the wetted parts of the valve can be eaten away and rendered useless or dangerous. Estimates of the annual cost of corrosion to industry are enormous.

The subject of materials is both a science and an art. Tremendous effort has been applied to laboratory and theoretical studies of metallurgy so that the chemistry of metal corrosion may be better understood and better metals and metal alloys thereby produced. Numberless field tests and experiments have been run. Plastics and ceramics have been studied and tried. Through theory and practical experience, the result has been improvements in both the understanding and application of materials for control valves as well as for other instruments and equipment. Where corrosion is bad but the use of the equipment is not otherwise severe, a valve body and trim may be made of plastic, or some parts may be fashioned of plastic-lined metal. Ceramics, too, have been used.

Yet the field results often fall short of those expected because of a lack of sufficient knowledge for a specific application. The subject can be a tricky one. A very small amount of an unexpected or ignored chemical contaminant may make a great difference in the corrosiveness of a chemical solution.

Sometimes a decision is made knowingly to forgo the use of the best material because of high cost—a compromise is made. Or the best material to resist corrosion may be too soft or have some other undesirable property.

6-1-5-3 EROSION

Erosion is another complicating factor. Sharp stones in a brook eventually become rounded by erosion caused by the gentle flow of water. The Appalachian Mountains were once as high and rugged as the younger Rocky Mountains are now, but they gradually eroded before the Rockies arose.

The same type of effect occurs in control valves except, because fluid velocities are high, in a greatly speeded-up way. We have seen that constricting a flow, as is done in a modulating valve, increases the velocity locally. The increased velocity aggravates the erosive action of the stream.

In the case of liquids, increasing the velocity may lower the liquid pressure sufficiently to create localized flashing, which means boiling (see Figure 3-17). If this happens, the resulting vapor bubbles increase the flowing volume within the valve, and the velocity increases further. There is then a mixture of liquid and bubbles of vapor zipping through the valve, with an intensified erosive effect. Flashing is damaging to the valve and often to the downstream piping.

There is more. We have seen how a liquid flowing downstream of a constriction recovers some of its pressure. If the piping downstream of the control valve is at a pressure above the vapor pressure (which is a measure of the tendency to boil at the existing temperature), then the vapor in the liquid-vapor mixture condenses and the bubbles collapse. This collapsing, known as *imploding*, causes powerful shock waves that tear at the valve materials and the nearby downstream piping. The phenomenon of flashing followed by implosion is called *cavitation*. If the downstream pressure is low enough to avoid creating implosions, then there is only flashing and the downstream flow remains as a mixture of liquid and vapor.

Cavitation is especially damaging. In extreme cases, within a few hours' operating time it will cause failure of some control valve components that are made of hard alloys. Cavitation is somewhat like sandblasting. It creates vibration and a sound like that of flowing gravel, and it pits metal badly. It is a problem not just for valves but also for pumps and ship propellers, which also create localized increases and decreases of liquid velocity.

The effects of cavitation may be minimized by choosing special types of valve bodies, including the *drag valve* and the *multistage valve*. A drag valve lowers the fluid pressure by passing the fluid through many narrow tortuous passages in parallel. A multistage valve takes the valve pressure drop in a number of steps in series. The proper choice of piping materials and the design of the layout may help protect the downstream piping against cavitation damage.

To minimize damage from all types of erosion, very tough and hard materials are required, especially for the valve trim. An example of hard trim is Stellite (manufactured by International Nickel Company). Where flashing and cavitation are expected, alloys other than ordinary carbon steel are required for the valve body, sometimes even for a short length of piping downstream of the valve.

6-1-6 Noise

Control valves can create audible noise that is sometimes unbearably loud. The fluid velocities in valves may be extremely high, and the amount of power dissipated by a valve can be very large. For example, consider a control valve handling 2000 gpm of cold water with a pressure drop of 1000 psi; this is not an extreme case. The hydraulic power to be dissipated is 1167 horsepower, which creates an extremely loud and harmful noise.

To protect the hearing of people who work in industrial environments, the Occupational Safety and Health Administration (OSHA) of the federal government has established standards for the maximum allowable noise. For example, a person is permitted to be exposed for eight hours a day to a sound level equivalent to that of an unmuffled truck engine. The permissible level of sound at the source increases as the person moves away from the source and as his exposure time is shortened.

Control valve noise is produced by the following:

1. The fluid flow.
2. Mechanical rattling, due to clearance between parts of the valve.
3. Natural frequency vibration, which is what a flagpole undergoes when there is a strong wind. The valve becomes a sort of tuning fork.
4. Valve position instability with resulting oscillation.

The control valve manufacturers can provide estimates of the noise their valves will produce in a given application. Proper selection of valve design, valve size, and actuator can reduce the noise.

To reduce the propagation of the source noise, the sound path has to be treated. This may involve the use of any or all of the following: heavy-wall pipe, acoustic insulation, and diffusers or mufflers.

6-2 Regulators

A *regulator* is a device that senses a process variable and automatically performs all the corrective actions needed to keep the value of the process variable nominally at a set point. The device does not require a separate outside power supply. In other words, it is a self-actuated closed control loop.

Regulators typically have only a nonadjustable proportional control mode, meaning that they cannot be tuned. Generally, they have a set-point adjustment. If more sophisticated control is required, then a regular controller and control valve are used. However, regulators fulfill many needs very satisfactorily and, in the smaller sizes, are generally less costly than a controller-control valve combination.

Examples of regulators for pressure, temperature, level, and flow are described in the next four sections.

6-2-1 Pressure Regulators

Figure 6-9 shows the basic designs of a pressure regulator, of which there are two basic types: the *reducing valve* and the *backpressure valve*, which control downstream and upstream pressures, respectively. In Figure 6-9, both types have the controlled pressure increase as the downward force of the spring is increased. Manually adjusting the spring adjusts the set point. Opposing the spring force is the upward force created by the controlled pressure acting upon the diaphragm.

These regulators are available for controlling gage pressure, absolute pressure, and differential pressure, depending on whether and how the diaphragm chamber of the actuator housing is pressurized (see Section 3-4-3).

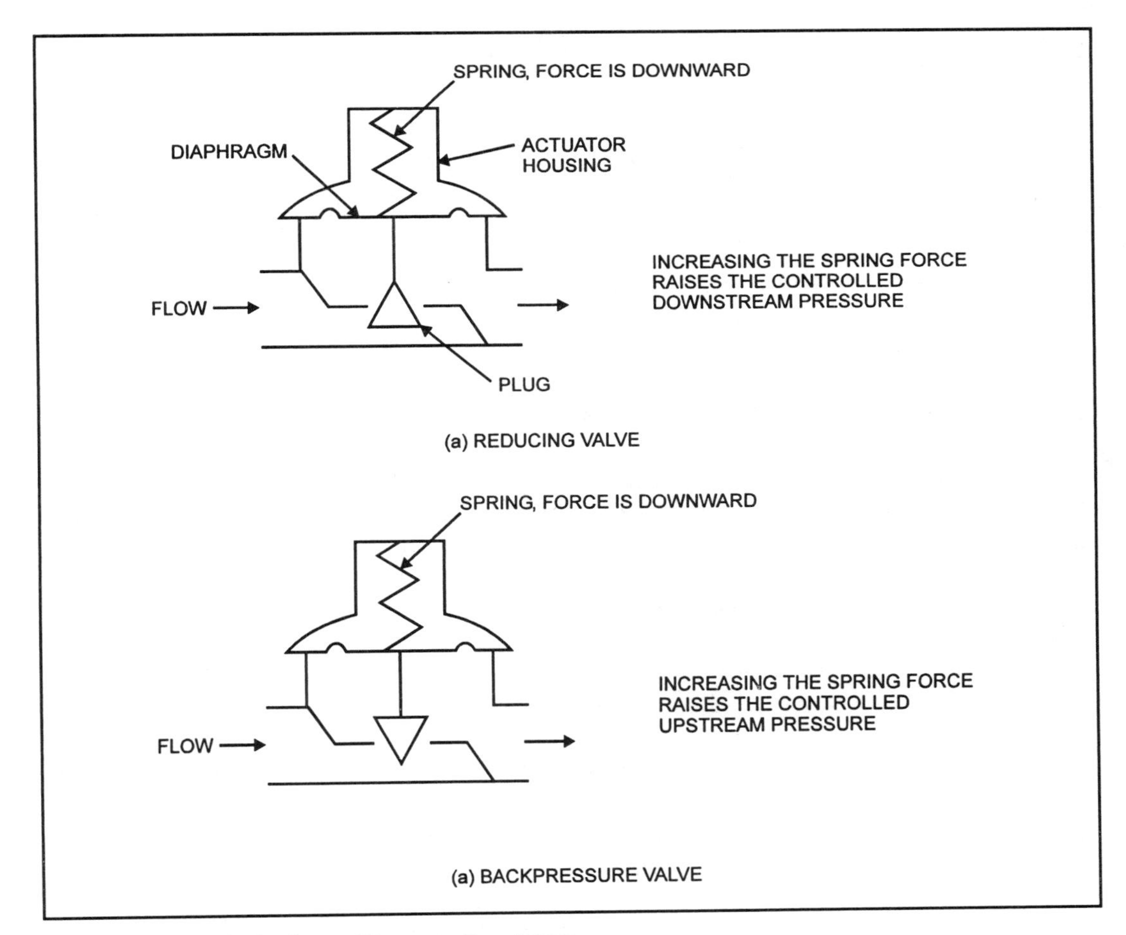

Figure 6-9. Basic Designs of Pressure Regulators

6-2-2 Temperature Regulators

A control valve that has a diaphragm actuator, as shown in Figure 6-6(a) and (b), uses a variable, controlled air pressure to modulate the valve. For a temperature regulator, the air pressure of the control valve is replaced by the pressure from a filled thermal system (see Section 3-7-2-4). The bulb of the filled system is connected through tubing to the valve actuator. As the sensed temperature increases, the pressure of the filled system increases, thereby moving the valve stem. The diaphragm force from the filled system is opposed by a spring. This arrangement is illustrated in Figure 6-10.

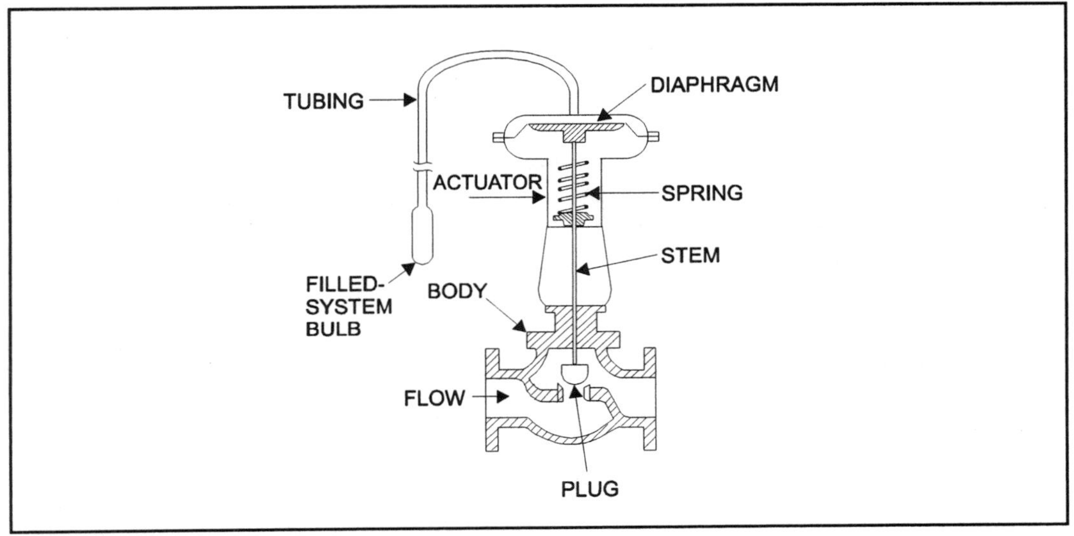

Figure 6-10. Basic Design of a Temperature Regulator

6-2-3 Level Regulators

The emptying of a toilet water tank is initiated manually. The tank has a level regulator consisting of a ball float that operates a water inlet valve to refill the tank automatically and then shut off.

A typical industrial level regulator has a ball float that rides up and down with liquid level changes in a tank. Through a mechanical linkage, the float movement causes a valve plug to modulate the flow to, or from, the tank. The regulator consists of the float, linkage, and valve, which are illustrated in Figure 6-11.

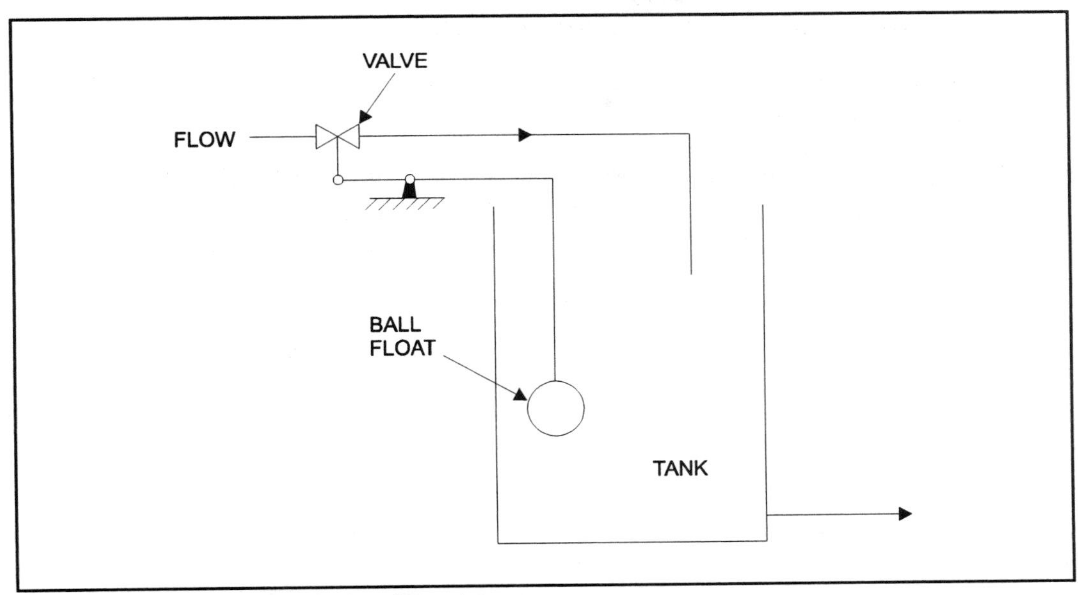

Figure 6-11. Basic Design of a Level Regulator

6-2-4 Flow Regulators

Flow regulators typically consist of a differential-pressure regulator combined in one assembly with a valve, which is simply an adjustable orifice. The basic principle of operation is diagrammed in Figure 6-12.

For a fluid at a fixed temperature and pressure, the flow through an orifice is constant if the pressure drop across the orifice is constant. For example, if we open our kitchen faucet, the water flow is constant provided that the city water supply pressure does not change. In Figure 6-12, the differential-pressure regulator maintains a constant pressure drop across the valve by modulating open or closed if either the inlet pressure or the outlet pressure changes; thereby the flow is kept constant. To increase or decrease the flow rate, the flow set point is changed by opening or closing the hand valve as required.

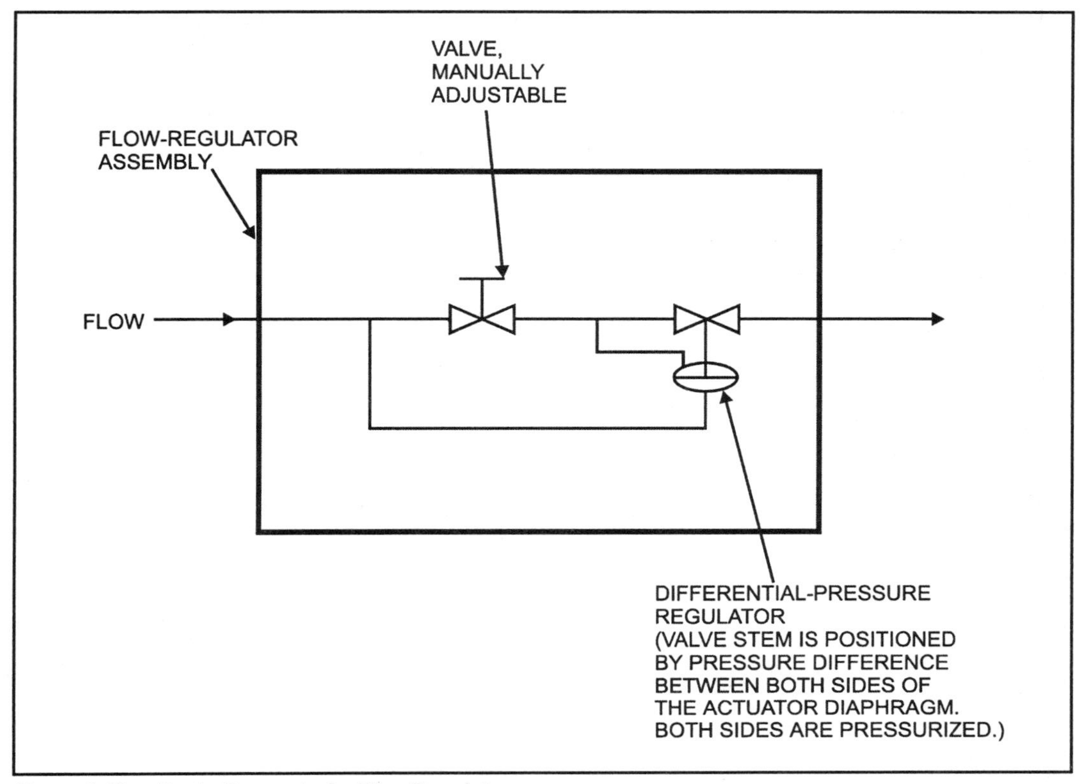

Figure 6-12. Basic Concept of a Flow Regulator

6-3 Limiting Elements

A *limiting element* prevents a process variable from exceeding a set limit. It is a device that senses a process variable and automatically takes all the corrective actions needed to prevent the value of the process variable from exceeding a set limit. Different elements are suitable for repetitive or for one-time-only operation. Examples of limiting elements for pressure, temperature, and flow are described in Sections 6-3-1, 6-3-2, and 6-3-3.

6-3-1 Pressure-Limiting Elements

6-3-1-1 Pressure Relief Valves

Safety valves, relief valves, and safety-relief valves are slightly different from each other, but their basic principle is the same. We will refer to them here simply as *pressure relief valves*.

These valves are used to prevent process equipment, piping, vessels, pumps, and the like. from bursting because of an excessive difference between the inside and outside pressures. If a tank is overpressured beyond its design rating, it bursts outward; it explodes. If a tank is under vacuum and has not been designed for it, it bursts inward; it implodes. A valve to prevent implosion is termed a *vacuum relief valve*. A typical relief valve to prevent explosion is illustrated in Figure 6-13.

The inlet of the pressure relief valve is piped to the equipment or system that is to be protected. In normal plant operation, the spring pushes the valve plug down against the process pressure, and the valve is closed. If all goes well, the valve will never be required to open during the life of the plant. But if the pressure were to become abnormally high because of control system malfunction, operator error, equipment failure, external fire, or chemical reaction, the valve would then open to limit the pressure buildup. This occurs when the process pressure against the bottom of the valve plug creates a force that exceeds the downward force of the spring at the spring set point plus the weight of the disk assembly. The spring force is adjustable to change the valve set point.

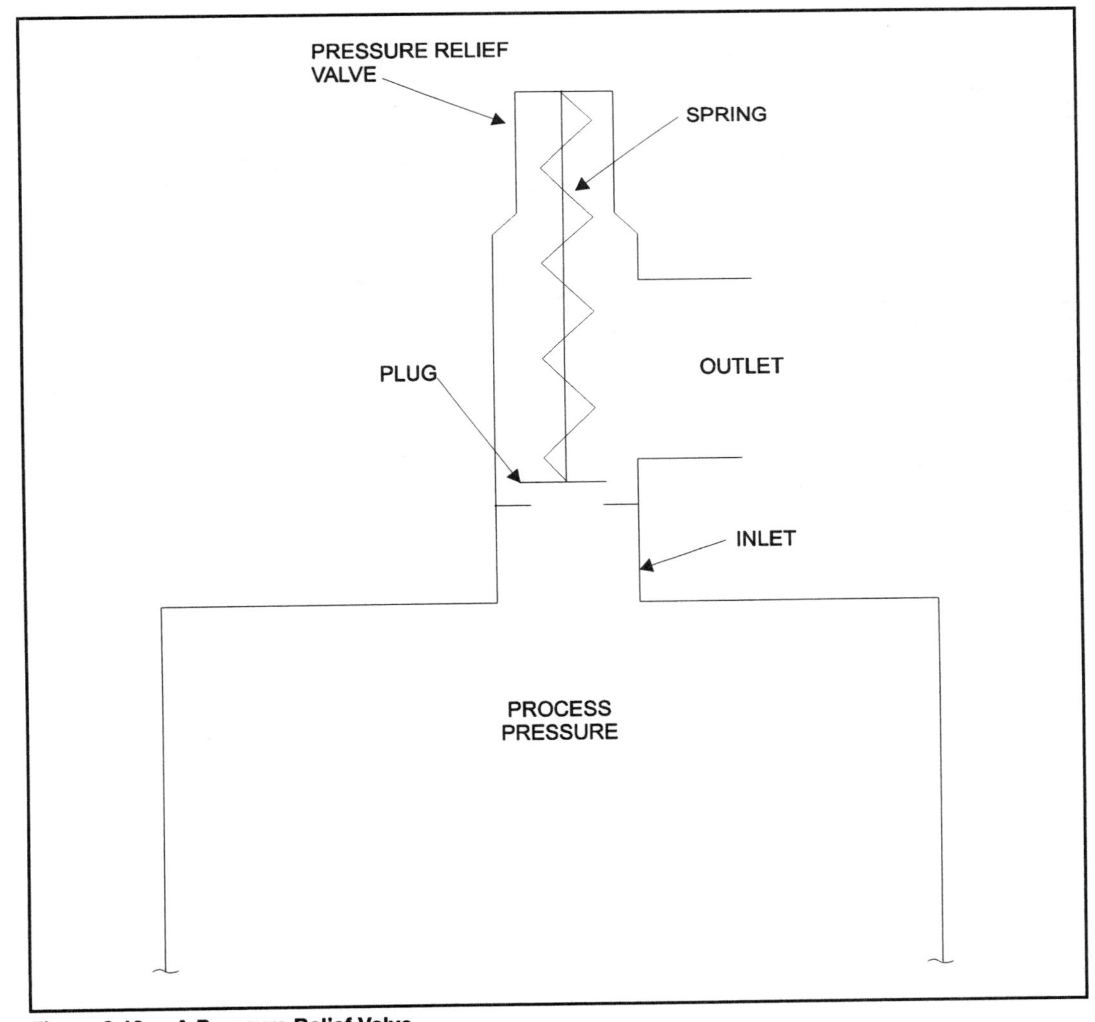

Figure 6-13. A Pressure Relief Valve

For the valve to open fully, the pressure must rise some predetermined amount above the set point. This incremental amount is known as *accumulation*, which is expressed as a percentage of the set point—typically, 3%, 10%, or other, depending on the application. A properly sized pressure relief valve, when wide open, limits the pressure rise to an acceptable safe value. A pressure relief valve is suitable for repetitive operation though it may develop leaks after it is operated, especially at high pressures. Mathematical formulas are used for sizing the valves according to the service conditions.

There are stringent standards established by the American Society of Mechanical Engineers (ASME), the American Petroleum Institute (API), and other engineering bodies regarding the design of equipment for pressure or vacuum service and the over- or underpressure protection requirements. These standards in many cases have quasi-legal status through the endorsement of governmental bodies.

The most widely used code for this subject in the United States is the ASME Boiler and Pressure Vessel Code, which covers the requirements for pressure relief valves. Many pressure relief valve manufacturers produce valves that bear an ASME nameplate attesting that the valves meet the ASME requirements. However, pressure relief valves are used wherever they are needed for safety, regardless of whether there is an explicit code requirement for the valve.

Both relief valves and backpressure regulators limit a rising pressure (see Section 6-3-1). Some pressure relief valves open proportionally with increasing pressure, as do many regulators. Others snap open. However, unlike pressure regulators, pressure relief valves are intended only to limit pressure, not control it.

The pressure relief valve codes are an essential point of difference between pressure relief valves and pressure regulators. The ASME code requirements can be met by pressure relief valves but generally not by regulators and generally not by an ordinary pressure control loop.

Good practice generally requires that there be at least two components to ensure pressure safety, as follows:

1. The first component consists of ordinary instruments to keep the pressure within safe bounds during plant operation. This component may be a pressure reducing valve or a control loop, which controls the inflow to the process system, or a backpressure valve or control loop, which controls the outflow.
2. The second component normally consists of a pressure relief valve (or a rupture disk—see Section 6-3-1-2) that should never have to perform if the first component does its job but stands ready to perform if the first component fails. This practice follows the principle of redundancy (see Section 11-2-3).

6-3-1-2 Rupture Disks

Rupture disks, also known as *safety heads*, are recognized by the ASME Boiler and Pressure Vessel Code. Like pressure relief valves, they are used to protect equipment against excessive pressure or vacuum. They are used either in place of or together with a pressure relief valve, depending on the application.

A disk is a thin diaphragm or plate that is usually mounted between a pair of pipe flanges and closes off the pipe. The disk may be made from a variety of metals, graphite, or other materials. Metal disks are sometimes plastic coated to protect the disk against thinning and premature rupture from chemical attack. Figure 6-14 shows a typical rupture-disk assembly.

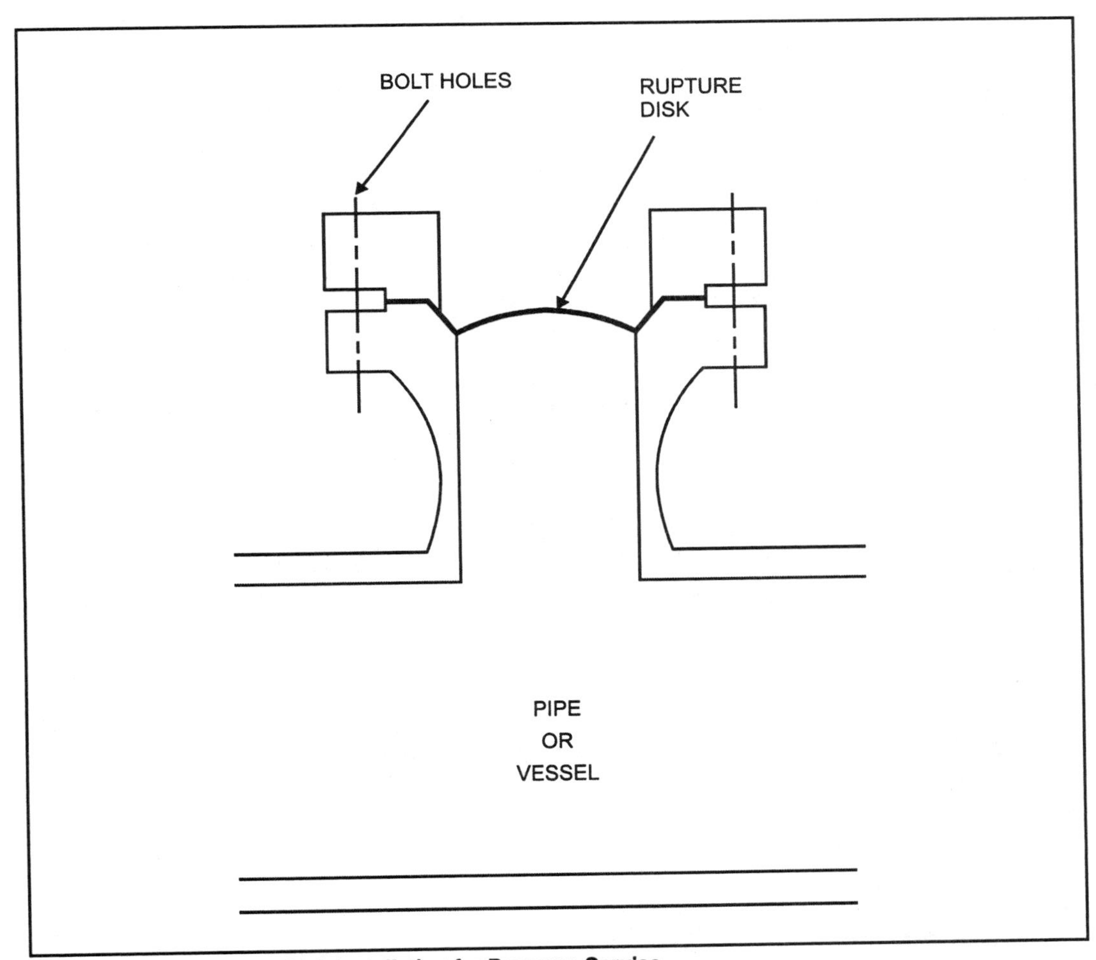

Figure 6-14. A Rupture Disk Installation for Pressure Service

The operation of a rupture disk is simple. When the process pressure rises to the design bursting pressure of the disk, the disk bursts, splits wide open, and the pressure is relieved. The design bursting pressure is equivalent to a set point but is not adjustable. Its value depends on the disk being properly manufactured and is, therefore, not completely accurate. In addition, metal disks may fail prematurely from metal fatigue. The lack of accuracy must be taken into account in designing the process equipment and the associated control system. In addition, if a relief valve and a rupture disk are for use in series, the valve may have to be derated unless the valve and disk are certified together. Nevertheless, rupture disks are very useful, especially when dealing with corrosive fluids or when all leakage from relief valves is to be avoided.

A ruptured disk is not reusable. Disks may be made of expensive metals that have salvage value.

6-3-2 Temperature-Limiting Elements

Fusible plugs or fusible links are temperature-sensitive devices that are used to limit temperature. A fusible plug may be a plug that normally blocks the outlet opening of a spray nozzle on a water sprinkler intended to fight a fire. Or a fusible part may take the form of a mechanical link that normally holds open a fire door that should be closed in case of fire. The plug or link is made from special metal alloys that melt at relatively low temperatures and enable the safety measures to perform—the sprinkler to spray water and the fire door to close.

Once used, the fusible element must be replaced.

6-3-3 Flow-Limiting Elements

An *excess-flow check valve* is a valve that permits a fluid to flow until the flow rate becomes excessively high, at which time the valve stops the flow. The cutoff flow is typically 1 1/2 to 2 times the normal flow. An excess-flow check valve is different from an ordinary check valve, which does not block forward flow but does block reverse flow. Figure 6-15 shows the basic design of an excess-flow check valve.

Reverse flow tends to close the valve, but a spring tends to keep the valve open. The spring is adjustable and provides the limit set point. As the flow increases, the closing force increases, and, at some point, the spring force is overcome and the valve closes.

Excess-flow check valves are used for safety against excessive flow caused by a broken line or other mishap. They can be used repeatedly.

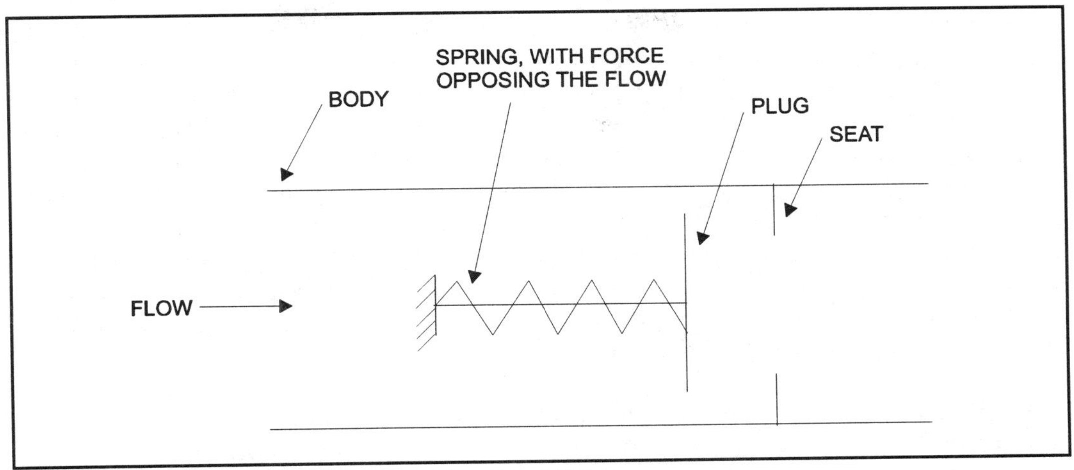

Figure 6-15. Basic Design of an Excess-Flow Check Valve

6-4 Other Final Control Elements

In previous sections, we saw how control systems use control valves as the final control elements to manipulate process variables. Other kinds of final control elements are presented in the next three sections.

6-4-1 Proportioning Pump

A *proportioning pump,* also known as a *metering pump,* pumps liquids by means of a piston or plunger inside a cylinder. Every back-and-forth stroke cycle of the pump discharges a fixed volume of liquid. The length of stroke of the pump can be adjusted manually, or remotely by using a pneumatic or other type of actuator. Each such adjustment changes the discharge volume per cycle.

Proportioning pumps are usually of low capacity and are frequently used to inject small volumes of a chemical solution into a relatively large stream. An example of such use is the chemical treatment of boiler feed water in a power plant to protect the steam generator against impurities in the feed water. A typical system uses a dissolved-oxygen analyzer, a controller, and a proportioning pump that injects hydrazine (a chemical) to lower the concentration of oxygen dissolved in the water. The pump is the final control element of a modulating analysis control loop. This is shown in simplified form in Figure 6-16.

6-4-2 Vane Control of Fans

Industrial fans may be supplied with adjustable *vanes, dampers,* or *louvers* for manipulating the gas flow through the fan. The vanes are placed on the inlet side of the fan and act, in effect, as a control valve. They can be actuated either automatically as part of a control loop or manually.

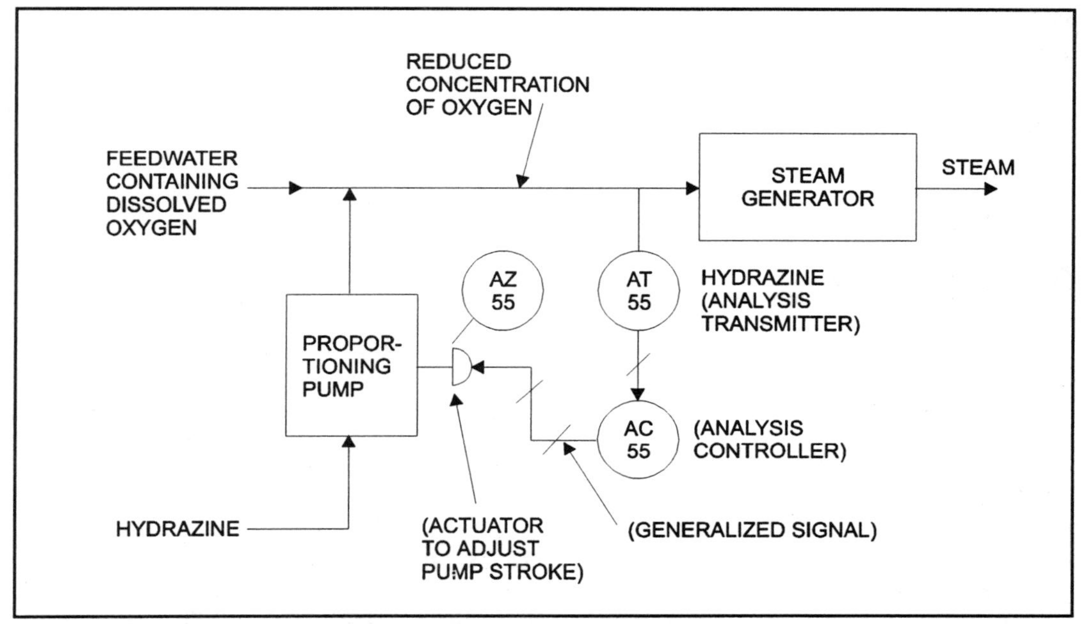

Figure 6-16. Using a Proportioning Pump as a Final Control Element

6-4-3 Variable-speed Control

The flow of materials can be modulated by varying the speed of process equipment that uses variable-speed drives. Examples are as follows:

1. Conveying solids by means of a variable-speed screw conveyor or conveyor belt.
2. Transporting liquids or gases in a pipe or duct by means of a variable-speed pump or compressor, respectively.

Power drives offering variable speed for process control include the following:

1. Electric motors.
2. Pneumatic motors.
3. Hydraulic motors.
4. Mechanical transmissions.
5. Steam turbines.

Variable-speed drives can be costly to purchase and maintain. The benefit they provide, however, is that they reduce power consumption by using no more power than is needed for pumping at any given time.

7

How Logic Control Is Performed

7-1 Definition of Binary Logic Control

Logic control is control that causes things to be done or not done to the process, depending on whether certain process conditions, operator actions, or control system actions happen or do not happen. Logic control makes use of a logic system, which is a combination of binary elements that automatically "decide" on the proper response to a certain set of conditions. The concept of logic control is outlined in Figure 7-1. (This traditional logic should not be confused with the entirely different "fuzzy logic" described in Section 5-1.)

In logic control, every part of the logic system—each input event, each logic function element, and each output response—is either *Yes* or *No*, *On* or *Off*, *True* or *False*, 1 or 0, as we saw in Section 3-2-2-1 on binary signals and in Section 4-2-1 on binary control. Logic control is appropriate for any process system that uses on-off devices to initiate or terminate normal or emergency operations.

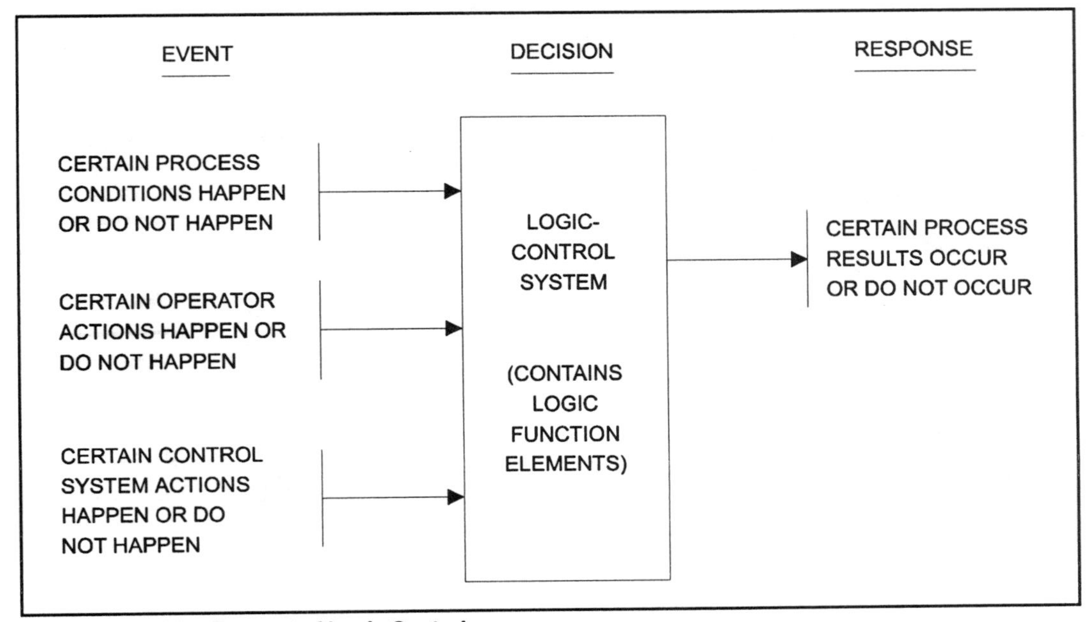

Figure 7-1. The Concept of Logic Control

7-2 How Binary Logic Control Works

Let's bring the definition of logic control in Section 7-1 down to earth. Some people we know are planning a picnic. Let's see how they do it.

Abner and Bella want to go on a picnic with Chauncey and Desiree. If they cannot all go this weekend, then they will go the following weekend. The process to be carried out is to get ready for the picnic and go. Figure 7-2 diagrams some of the possibilities. The symbolism used is taken from ISA Standard S5.2, *Binary Logic Diagrams for Process Operations.*

Figure 7-2 is a logic diagram that shows the following requirements that need to be satisfied in order for the group to go on the picnic this weekend:

1. Nobody is sick.
2. Either (a) Abner does not get an emergency call to work this weekend or (b) he gets the call but is able to get Zorro to fill in for him.
3. Food and drink are prepared.
4. A car is available.
5. The weather forecast is good.

If every one of these requirements is met, then the group goes on the picnic this weekend. If a single one or more of the requirements is not met, then they must wait a week.

The AND function in Figure 7-2 requires all its inputs to exist and continue to exist in order to obtain and maintain a logic output. If any input ceases, the AND output ceases. The OR function requires only that one or more of its logic inputs exist in order to have and maintain a logic output. The circle represents the NOT function, which really means IF NOT: if there is no logic input, then there is a logic output, and if there is a logic input, then there is no logic output. The logic functions are sometimes called *gates*; for example, "an AND gate."

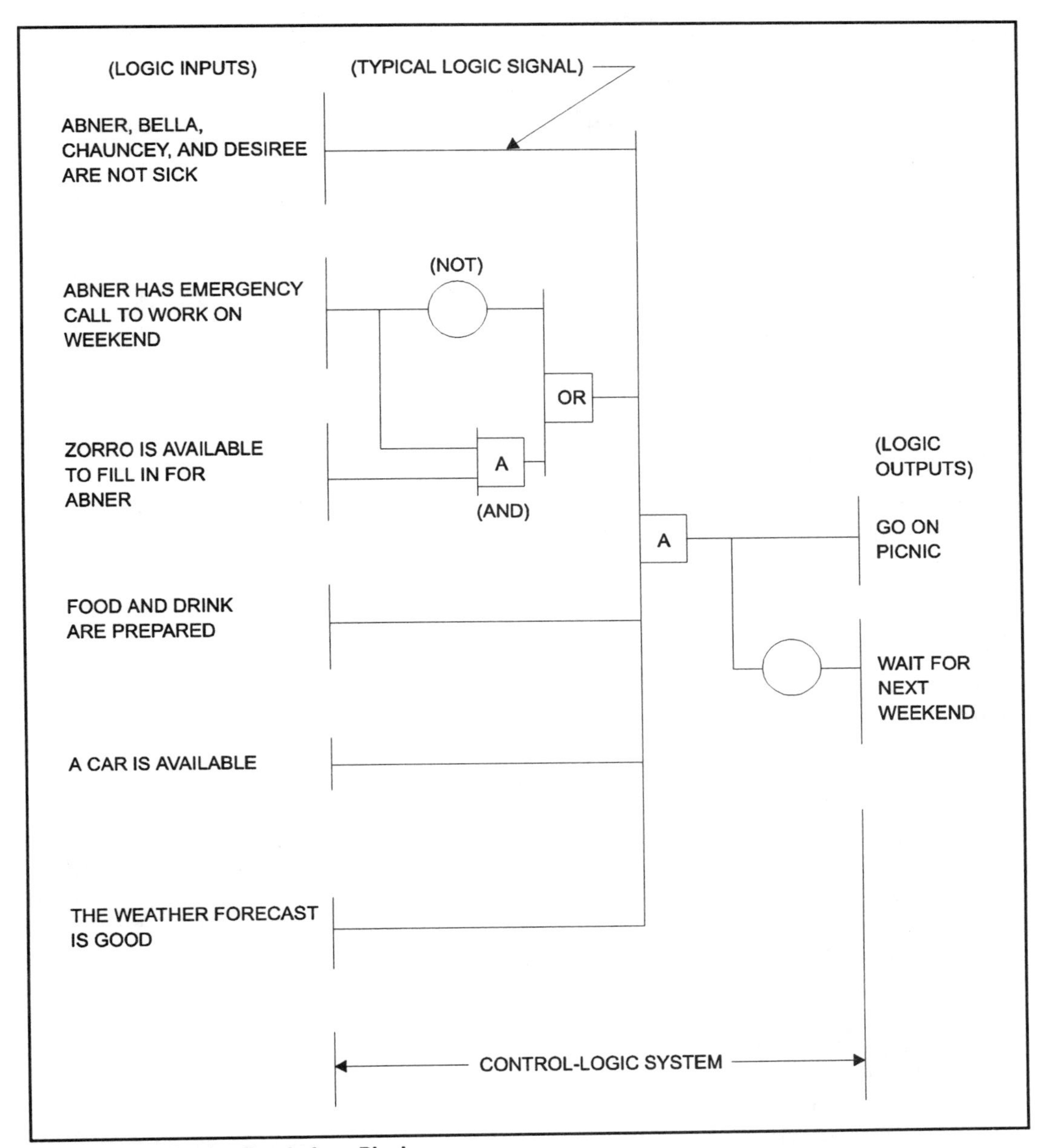

Figure 7-2. Getting Ready for a Picnic

In the weekend picnic example, the logic signals and function symbols exist only in our minds. But for a process plant, a logic diagram uses actual hardware to carry out the logic. Consider the following process plant situation. A plant has a chemical pump that is started by push button. The pump requires a supply of lubricating oil before it starts. The supply could be turned on manually, but, for the sake of reliability, it is decided to do it automatically. So the following must first occur: A lubricating-oil pump must start automatically when the chemical pump *Start* push button is operated, and normal oil pressure must be established within six seconds. The operation logic sequence is as follows:

1. If the oil pressure is normal—not low—after six seconds the chemical pump operates. Otherwise, the startup is cancelled automatically.
2. The operator can stop the operation at any time by pressing the *Stop* push button. Independently, the operation stops if the oil pressure becomes low.
3. When the chemical pump is commanded to stop, its power is cut off immediately and it stops, but the oil pump should continue to operate for ten seconds, then stop. The 10-second delay ensures that the chemical pump continues to be lubricated while it is slowing down.
4. While the chemical pump is required to operate, an alarm, PAL-64, is actuated if the oil pressure becomes low.

Figure 7-3(a) diagrams the logic for starting and stopping the chemical pump normally and for protecting the pump against low oil pressure during operation. We should remember that the operating lines in the figure represent logic signals, not physical signals.

Now consider the case in which the system is down and there is no oil pressure, which is a normal situation when the chemical pump is not required to operate. A low-oil-pressure alarm is triggered, which is false because there is no trouble. Nevertheless, the alarm requires an acknowledgment action on the part of the operator, and the alarm light remains on because the apparent trouble, low oil pressure, persists.

Such a false alarm is known as a *nuisance alarm,* which exists here because the system logic does not recognize that the system is in a shutdown state. Nuisance alarms are an inconvenience and, worse, a distraction for the operator (see Section 11-3-2, "Presentation of Information").

A nuisance alarm can be eliminated by permitting it to be actuated only in the event of a genuine problem or abnormality. In our example, the alarm logic should be as follows: while the chemical pump is required to operate, an alarm is actuated if the oil pressure becomes low provided that six seconds have elapsed after the *Start* button is pressed. Thus, the oil pump operates and has time to build up the oil pressure before the alarm can operate. Figure 7-3(b) shows the logic for the pump system with the nuisance alarm blocked.

All the stated logic conditions and actions are binary in nature. The oil pressure is either low or not low, which is signaled by the output from the on-off switch, PSL-64. The time-delay elements have an output or not, depending on whether the specified time has expired or not, the pumps operate or not, and so on.

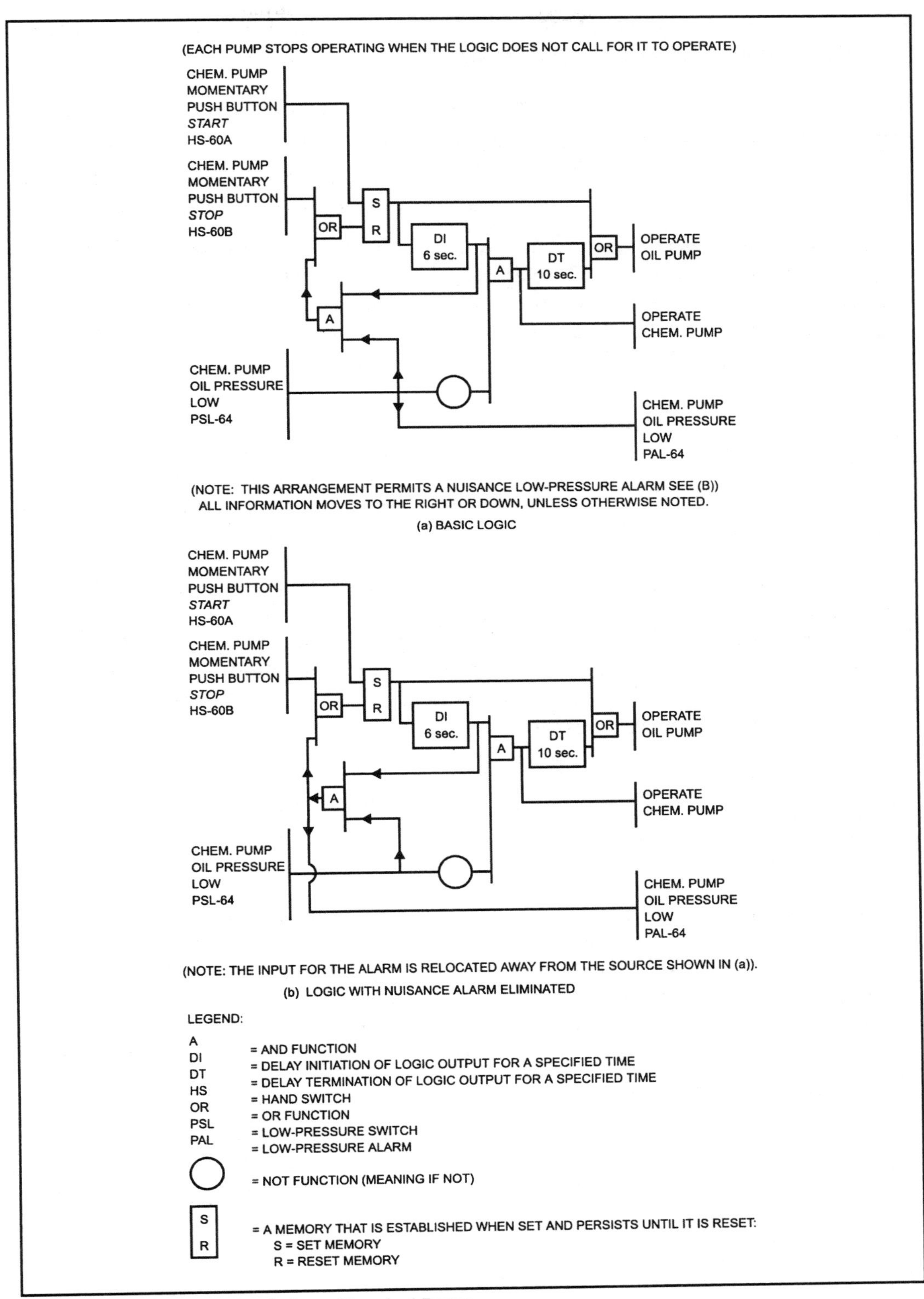

Figure 7-3. Operating Logic for a Chemical Pump

A different situation may exist where there is an analog control loop whose action should be blocked under some circumstance. For example, Figure 7-4(a) shows a modulating flow control loop controlling the flow of liquid to a tank. The outflow from the tank is controlled by a level controller. No excessively high level is expected, but if it were to occur through some mishap, the flow to the tank must stop.

The flow controller signal passes through a small solenoid-actuated, three-way pilot valve that normally does not hinder the signal to the control valve. The controller signal is pneumatic, 3 to 15 psig. But if the tank level becomes abnormally high, the high-level switch output shifts from *On* to *Off*, the pilot valve moves to vent the control valve actuator, the control valve fails closed on loss of control air, and the flow to the tank is stopped.

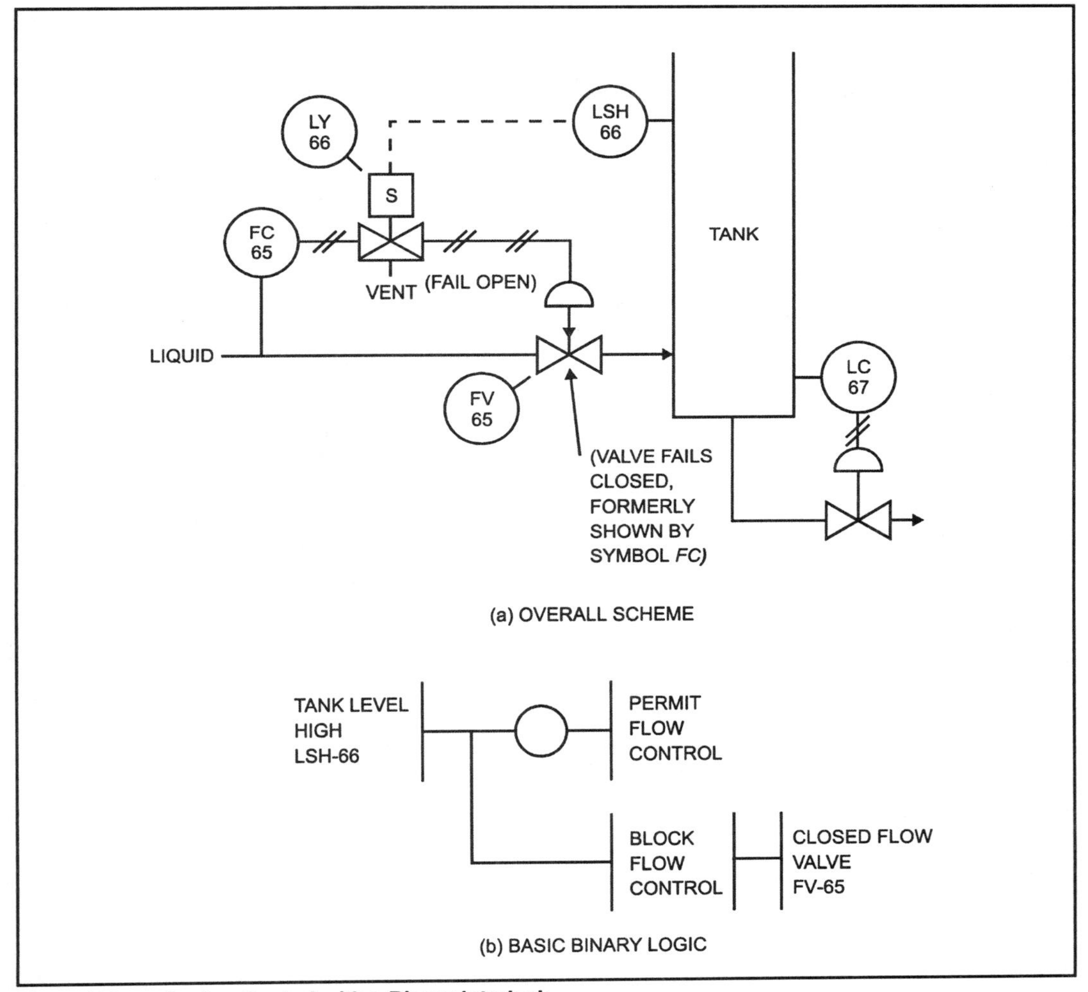

Figure 7-4. Analog Control with a Binary Interlock

What we have here is a binary control loop—the level switch and its pilot valve—that overrides an analog controller. The basic logic is shown in Figure 7-4(b). The *override,* also known as an *interlock,* disrupts the process operation by stopping the flow to the tank. Compare this with the selective control of Section 4-3-3, which has a mutual override of two analog controllers with no interruption of the process. Each scheme has its place.

A kitchen example of a safety interlock is a microwave oven that cannot operate until the door is closed.

7-3 The Advantages of Binary Logic Control Diagramming

Logic diagramming arose in response to a need for planning logic systems that were far more complex than the system shown in Figure 7-3. Written or spoken verbal descriptions of interlocking process operations are often difficult to piece together mentally to provide a whole picture. Logic gaps or errors may go undetected.

The logic control plan should be correct and clear before we attempt to detail the hardware circuits to carry out the plan. Otherwise, we risk getting on a horse and charging, so to speak, before we know where we are going. Let's do it right the first time, the zero-defect way.

The few symbols of the logic language together with some English-language words enable a coherent and explicit description of the operating logic to be diagrammed. After a little familiarization, the graphic nature of the diagram makes it relatively easy to read and understand. Logic inconsistencies become easier to catch. The method is simple in that it focuses on the process operating requirements, not the details of any particular kind of hardware. Therefore, the logic depends, for example, on whether a flow is high or low, not on whether the electrical contacts of the flow switch are open or closed. The mechanical details of the switch action are left for the technical specialist.

At the same time, the logic diagramming can present the operating requirements with a greater degree of detail depending on for whom the diagram is intended. The level of detail may range from that required for a designer of detailed control circuits to that required for a project engineer who needs some understanding of the operation of the plant that his subordinates are working on. Other types of people who may get involved include process engineers, equipment engineers, plant operators, and service personnel.

7-4 Implementation of Binary Logic Control

7-4-1 Logic Circuit Components

All logic control functions can be carried out using several different types of hardware, either unmixed or mixed. The hardware consists of individual components that are connected together by wire conductors, printed electrical circuits, tubing, or optical fibers to form a logic control system, as shown in Figure 7-1. The components of a system may include the following types:

- *Electromechanical Relays*. These relays are of the traditional electromechanical types that consist of an electrical solenoid that, when energized or de-energized, mechanically moves one or more electrical switch contacts to open or close electrical circuits.

- *Electrical Solid-state Devices*. These arose following the invention of the electrical transistor, and they now hold a dominant position. They perform switching operations but have no mechanical movement. They are steadily supplanting electromechanical relays as control elements but, generally, have a smaller current-carrying capacity than do electromechanical relays.

- *Fluid-power Devices*. These are usually pneumomechanical devices, essentially very small air-actuated valves and power pistons. They are very widely used for controlling machine operations in manufacturing plants. Pneumomechanical devices have limited application in process plants, usually in small, field-mounted systems. The larger systems are normally electrical.

- *Other.* These include fluidic devices, which use air but have no moving parts; hydromechanical devices; optical devices; and manual devices.

7-4-2 A Working System

An electrical schematic, or equivalent, is the definitive diagram for showing how the operating requirements are to be, or have been, carried out. It reflects the circuit designer's knowledge of the intricacies of control circuits and their hardware. Incompatibilities that may exist between the logic diagram and the actual circuit must be ironed out before either diagram can be considered final.

An engineering diagram is required if electromechanical relays will be used. Such a diagram is called a *ladder diagram, relay ladder diagram,* or *electrical schematic diagram* (see Figure 7-5; also see the discussion in Section 8-3-2-3, distributed control systems [DCSs] and programmable logic controllers [PLCs].)

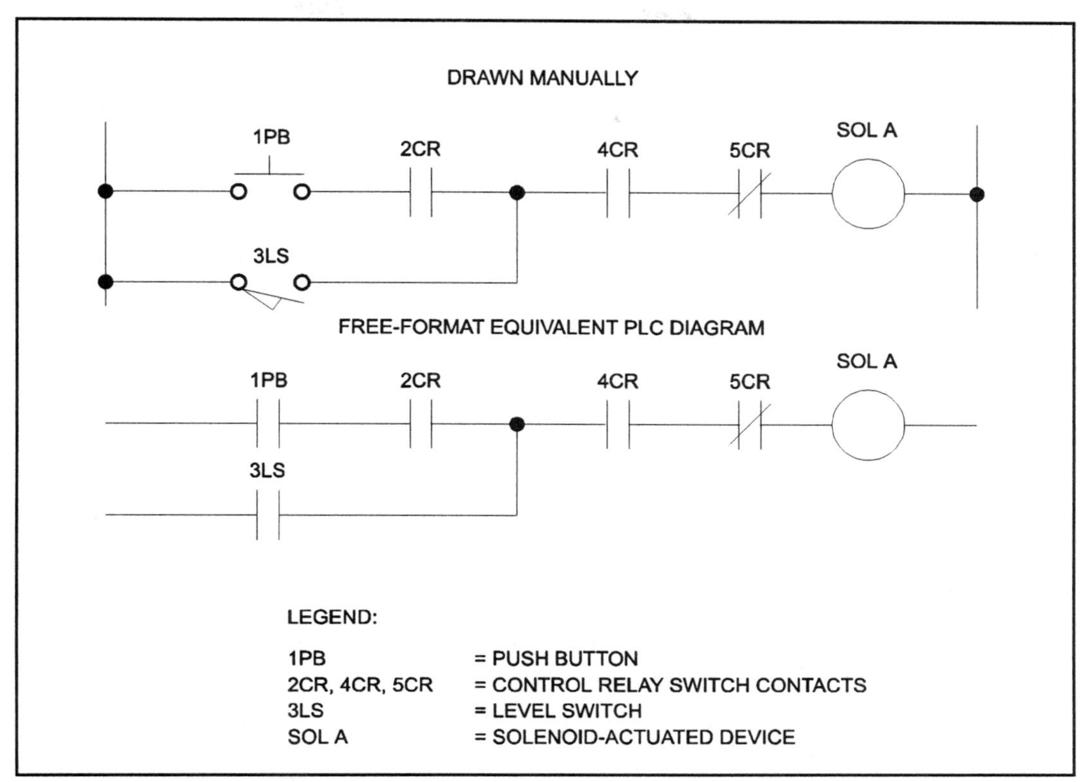

Figure 7-5. A Relay Ladder Diagram

8

THE COMPANY HIERARCHY OF CONTROL

8-1 ACHIEVING THE COMPANY'S PRIMARY GOAL

So far we have focused on the plant process: how to measure what is going on and how to control it. However, the plant that is supposed to operate efficiently is only one element of a grander operating system of a different kind. The larger system is the plant owner's whole organization, whose overall purpose is to be run as a profitable business by satisfying a public demand for the plant's product. This is why the owner invested his effort, time, and money in establishing the plant. The production plant is only a means to an end.

For the company, the profit-making process is the production and sale of goods, which is part of a closed-loop control system, as shown in Figure 8-1. This loop is an application of the generalized closed control loop of Figure 4-1. The profit goal is the control set point, which really is a minimum goal: the owner does not object to overshooting the goal, but he would rather not fall short.

The production portion of the process involves obtaining and using information concerning the manufacturing process. Some of this information—on production, inventories, and so forth—is used also by upper-level managers for planning.

8-2 THE CONTROL HIERARCHY

A company's control of its plant's operation has several layers of responsibility—a hierarchy of control. There are no hard-and-fast categories of control that fit the needs of different plants and different companies. However, a representative arrangement may be as shown in Figure 8-2. This figure outlines the general plan and information relationships for three ranks of authority and responsibility in a company having several production plants.

The ranks of control, beginning with the highest, are as follows (also see Section 8-3-2-2 on distributed control):

- Rank 3 is the central management, which plans the master strategy for the company operations. Its people may be likened to the general staff in the army. The management is concerned about the balance between production and sales, among other things. For example, the management

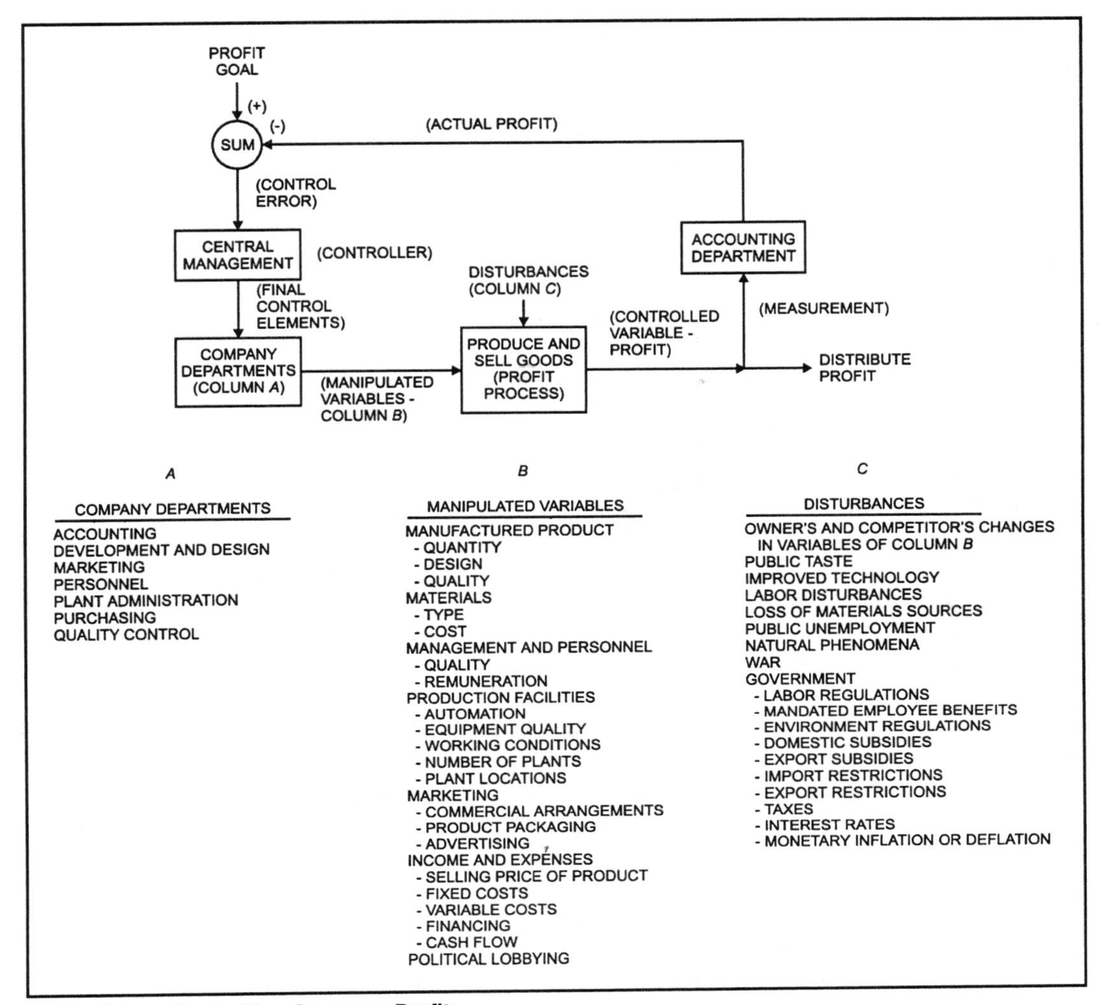

Figure 8-1. Controlling Company Profit

of a petroleum refining company is informed that the coming winter is expected to be unusually cold. Their planners decide that they want to produce more heating oil and less gasoline. Based on analyses of the performance, capabilities, and inventories for each of their oil refineries, the planners specify the product mix they would like to see from each refinery in order to optimize the performance of the company as a whole. The responsibility of each plant for meeting the production target falls to Rank 2.

- Rank 2 covers the plant management for each refinery. The management plans the strategy to optimize the operations so the refinery may meet its production goals most efficiently. They establish the operating requirements for the refinery. This management is like the commissioned officers of the army who plan the tactics to achieve the military goal they have been given by Rank 3.

- Rank 1 consists of the operating people who directly control the process in their particular plant. These people are like frontline junior officers and enlisted men. They carry out the tactics for victory as directed by Rank 2.

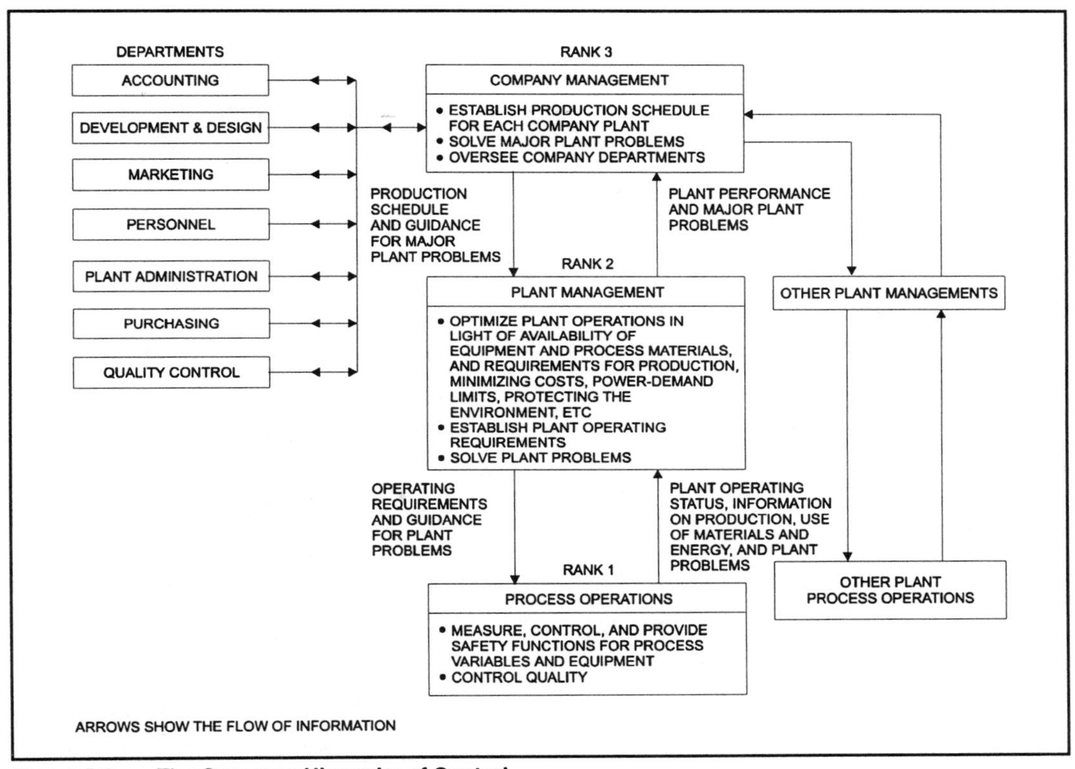

Figure 8-2. The Company Hierarchy of Control

Rank 1 makes use of process instrument systems for measurement and control. The control systems may include the following types, which are discussed in Section 8-3:

1. Traditional field control.
2. Traditional centralized control.
3. Computer control, leading to
   a. Supervisory control.
   b. Direct digital control (DDC).
   c. Distributed control.

8-3 The Format of Process Control Loops

The measurement and control systems that are used by Rank 1 for direct control of plant operations have different formats, that is, different arrangements, for separating or grouping loop or instrument functions in different plant locations. The word *architecture* may be used instead of *format*. The format of the systems is outlined in the following sections. However, the field of instrumentation is in flux, with constant mutation and hybridization of basic instruments and their systems. It is difficult to cover the field broadly without finding exceptions to some statements in the descriptions. However, the descriptions do point out practices and possibilities.

8-3-1 Traditional Instrumentation

8-3-1-1 Field Control

The domain of process instrumentation began with only field control, which remains important now but not exclusively so. Furthermore, field control will continue to be used for many services in the future.

Field control has all the instruments of a loop locally mounted, generally near the related plant equipment. Field control provides operating information right at hand for operators who are stationed there or who make a routine once-a-day or once-a-shift inspections of the plant equipment and operation as well as for people who service the equipment.

Field control is also the simplest system. It has the fewest links to break in the instrument chain, thus enhancing system reliability. It avoids the cost of running signals from a local point in the plant to a remote control center.

Such control is perfectly appropriate today in many cases for control loops that have a constant set point, as is often true for plant utility systems. For example, say a plant produces steam at 300 psig but requires that some steam be used at 15 psig. A local pressure control loop that is set unvaryingly to control at 15 psig will often be the correct type to use. There is no need for operator adjustment, hence no need for the controller to be in a control center for frequent manipulation. (There may be a remote alarm system to guard against the consequences of a failure of the local control system. The very earliest alarms were also local.)

Figure 8-3 illustrates the arrangement for direct-connected instruments, showing both how they are used and the system format. The control is totally local and requires a local operator to carry out the Rank 1 responsibility. Ranks 2 and 3 have no tie-in to Rank 1 other than through electronic, written, or oral communication.

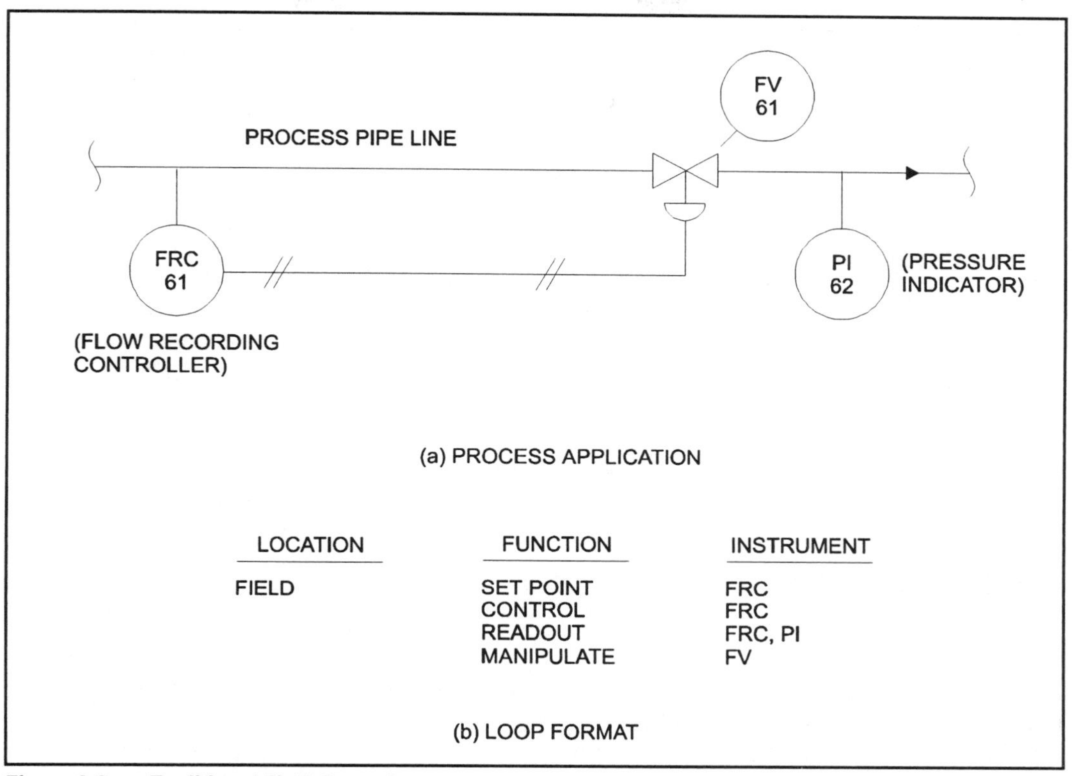

Figure 8-3. Traditional Field Control

8-3-1-2 Centralized Control

The locations of local control loops in the traditional field arrangement were widely scattered. To reduce manpower requirements and improve the coordination of operations of the various parts of the plant, instruments became less isolated, beginning with local groupings and culminating in a control room that now is the control center for the entire plant. The control room is designed so that all important operations can be given close surveillance and can be carried out from this location.

The centralized instruments were originally mounted in rows on instrument boards of steel or other material. The boards were at first made of vertical panels approximately seven feet high, placed side by side in a straight line. Then the boards became U-shaped because of their great length (up to 50 feet or more) and the vertical panels developed sloping segments at the top and at bench height for the convenience of the operator and to improve the readability of the instruments. (It was claimed but never authenticated that the operators needed roller skates to cover some of the large boards.)

Instruments used to be large, approximately fifteen inches, or more, in height and width. In the late 1940s, miniature indicating and recording controllers, six inches wide by six inches high—and later in widths down to one inch—were introduced.

These permitted closer spacing of instruments and the placement of a greater number of instruments on the central board. These dimension changes took place initially for pneumatic instruments, then for electronic instruments. This concentration of instruments was accompanied by the replacement of relatively bulky alarm lights by annunciator cabinets that had windows that utilized board space more efficiently.

The additional instruments on the board increased the amount of process information that the operator had to deal with. At the same time, the miniaturization of instruments helped to make practical the *graphic panel*, which was often used to improve the operator's command of the process. A graphic panel, sometimes called a *mimic panel*, is a panel bearing a simplified diagram of a plant process or a portion of it. The diagram is usually made from flat plastic pieces; color coding is used to make the diagram easier to follow. The control instruments, readout instruments, and lights are placed in the diagram where they fit in the process. ISA standard S5.5, *Graphic Symbols for Process Displays*, has established standard graphic symbols for use in process displays on both graphic panels and video screens.

Another development to help the operator was the control console, which was a low instrument board with a projecting shelf. The control console was to be roughly U-shaped, patterned after an organ console. The theory was that a seated virtuoso operator, by masterful and deft movements of his arms and fingers, would be able to reach and operate all the instruments needed to make the process plant play in harmony (even without the benefit of bass pedals). An air pilot's cockpit has also been fancied as the prototype for the console.

As consoles developed, they remained faithful to their name but, unfortunately, not always to their intent: An operator would often need an extensible arm or the long arm of a gorilla to perform on his or her instrument while remaining seated. Nevertheless, consoles have eased the operator's job considerably. This is also true for the distributed control systems that have followed.

Data loggers, which are printers of as many as several hundred or more pieces of process data, often with alarm, calculation, and simple control capability, began to appear around 1960. In the 1970s, data loggers began to sprout video terminals with keyboards. These were essentially improved devices for doing the same old things but with more flair. They did not cause a basic change in the physical format of control or of control centers.

The gathering together of instruments in the control room requires transmitters to bring the process measurements from the field to the control room instead of having process sensing performed directly by the remote controller or indicator. It is generally more practical, safer, and less expensive to communicate by pneumatic or electric signals over long distances, even within a plant, than to run sensing lines over those distances.

Figure 8-4 shows two variations of traditional centralized control loops. Part (a) of Figure 8-4 uses a conventional, self-contained controller, which contains two groups of functions, which illustrate the issue of format, are as follows:

1. *Control functions*. This group includes (a) the control modes—proportional, integral, and derivative—and (b) the configuration adjustments for tuning the control modes and for choosing direct or reverse action. An instrument that contains these functions is a *controller.* It may also contain other functions.
2. *Operator functions.* This group (a) indicates the operating variables—process measurement, controller output, and often the deviation from the set point and a deviation alarm—and (b) contains the operator's adjustments for the set point and the selection between automatic and manual control. An instrument that contains the operator functions and does not include the control functions is designated a *control station.*

All controllers used to contain both control and operator functions. With miniature instruments, which initially were pneumatic, came the frequent separation of these functions into what is known as *split architecture.* The purpose of the separation was to permit the control function to be located close to the process in the field while retaining the operator's command in the control room. This was done to eliminate the pneumatic transmission time lags to and from the control room, thereby improving control significantly for fast loops—typically, pressure control and flow control.

Most control room controllers today are electronic and have essentially no signal transmission lag. However, split architecture continues to be used, particularly for large plants, for the following reasons:

1. Relocating the configuration hardware permits the control station to be made smaller, resulting in a more compact instrument board.
2. The control room operator may not need access to the configuration adjustments if these are left for an instrument technician to handle elsewhere.

Figure 8-4(b) uses the control scheme of Figure 8-4(a) except with split architecture. If the remote controller is pneumatic, its location is near the control valve; if electronic, the location is in an equipment room or other convenient place.

The format of traditional centralized control fulfills the requirements of Rank 1 and also provides a more comfortable environment for the control room operators. This format tends to reduce the number of plant operators needed and simplifies the job of running the plant as a unit rather than as a collection of disjointed parts.

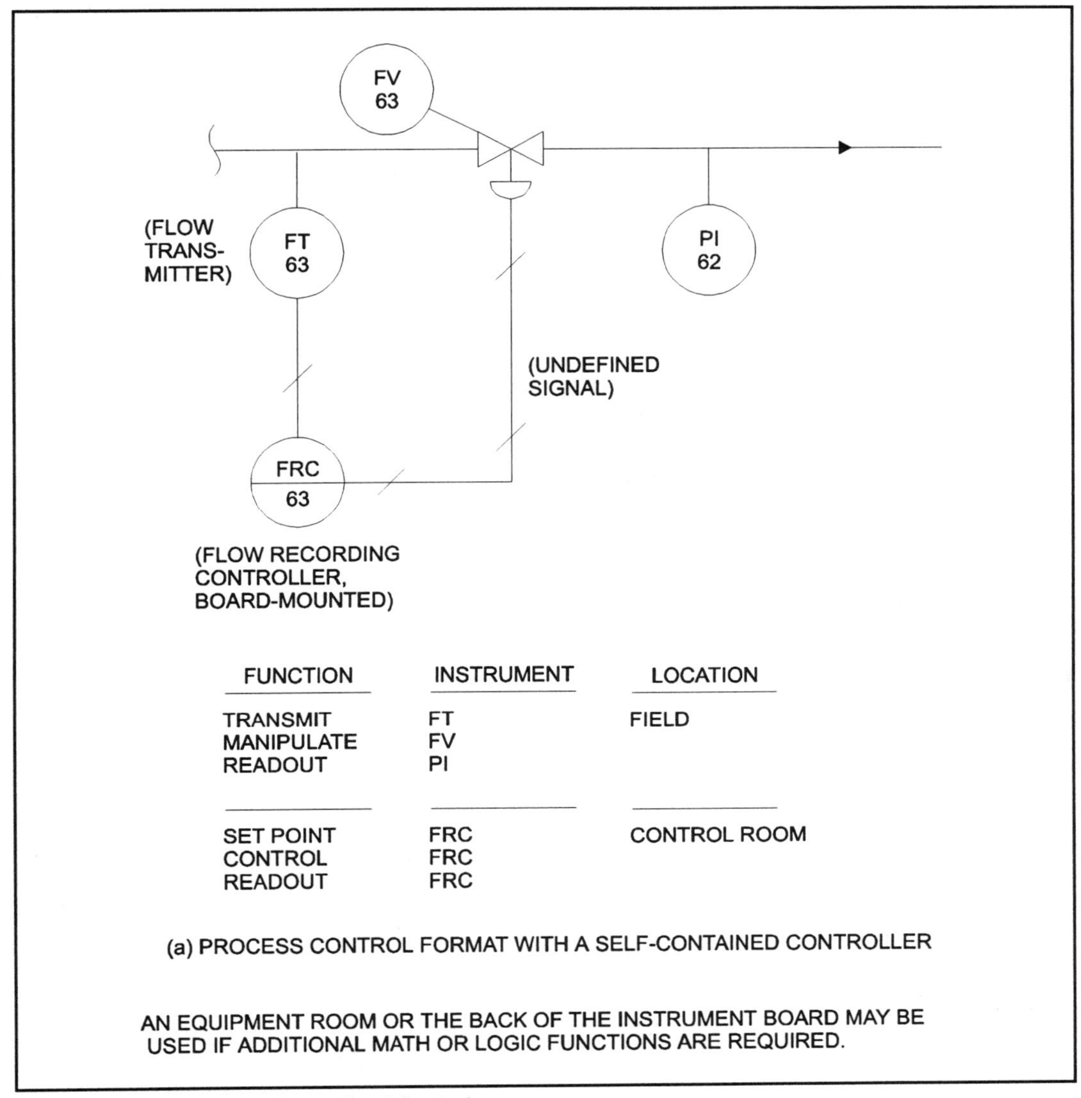

Figure 8-4(a). Traditional Centralized Control

With Rank 1 having centralized operating information and the capability to gather and report information by means of data loggers and other printing devices, Rank 2 management has easy access to data on key operating variables in the plant and can be kept more up to the minute regarding plant status, than in the past. Rank 2 may use a computer to derive information, such as plant efficiency, and to communicate with Rank 3.

Generally, the operation of Rank 3 is not much changed and depends largely on plant information obtained from Rank 2.

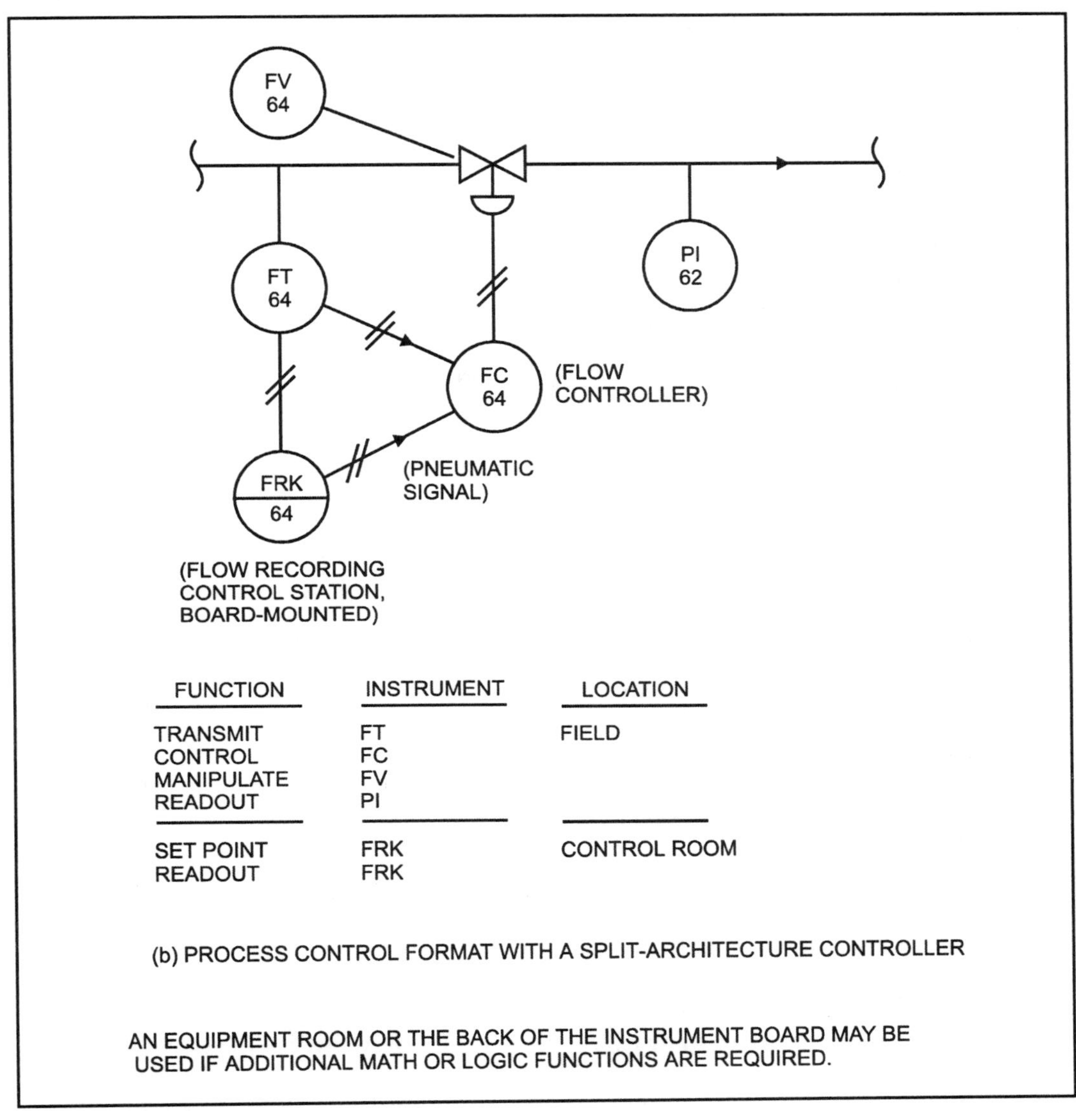

Figure 8-4(b). Split-Architecture Control

8-3-2 The Microprocessor Revolution

A *microprocessor* is a small electronic computing element that can contain the equivalent of millions of electrical circuits. Microprocessors, which have brought on the Second Industrial Revolution, are continually increasing their capabilities at an explosive rate.

The revolution began with the invention of the transistor in 1947 by William Shockley, together with John Bardeen and Walter Brattain of the Bell Telephone Company. This led to the microprocessor that has transformed commonplace objects: wristwatches, cameras, kitchen ovens, calculators, and innumerable others. In process control, the microprocessor has improved individual instruments and made them more versatile, and it has caused radical changes in control systems. The systems come in almost endless variations, of which representative examples are described in the following sections.

8-3-2-1 Computer Control

The computer has become a major player in the game of process control. The computer was first used as a large machine that simply gathered and recorded data. It was a high-class multipoint recorder (see discussion of data loggers in Section 3-3-2-1 on "Chart-type Recorders").

The next step was to use the large, or mainframe, computer for *supervisory control,* or *set-point control,* in which the computer performed control computations and sent a set-point signal to an analog controller, which would do the actual controlling.

The large computer was next used for *direct digital control* (DDC), which bypassed the intermediate analog controller and operated the final control element directly. The analog controller could be retained as a standby in case the computer broke down. This system had an operating program with the potential to model the process, optimize it, and provide nonstandard control functions. It seemed like the answer to many problems.

But DDC based on the mainframe computer never went far. The computer was a multipurpose machine that needed custom programming to instruct the computer in (a) how to become a controller and (b) what it should do after it became a controller.

This was comparable to your hiring me and sending me to a school for butlers so that I would learn how a butler behaves and also learn the skills of a butler. I would be programmed to "buttle." After I earned my butler's diploma, you would rent me out to parties. For each party, someone would configure me so that I would know whom to announce on his or her arrival, where the dishes were kept, when dinner was to be served, and so on. I would be reconfigured for the next party, but my programming would not change.

In the early days, the application skills for programming a computer for process control hardly existed. It was a forced marriage, so to speak, and close cooperation was required between control engineers who did not know how to program a computer and computer programmers who did not know processes or process control—a bad match. In each case, the programmed product was one of a kind; it was expensive, slow to prepare, and slow to be revised. Another negative factor was that many present-day reliability techniques for electronics and sophisticated control systems were not then established. The risk of shutting down an entire process because of a failed part in the single controlling computer was generally unacceptable.

Computers have progressed so that small, modular, standard controllers are now available. However, they may not be standardized among manufacturers. A major difference between the old and the new computers is that the old required programming and configuration, and the new require only configuration. The manufacturer now builds in the programs for control modes, calculations, logic, and the like, which are called *algorithms* or *firmware*. The controller already knows how to control. All that the controller now lacks is a knowledge of the specific adjustments for each application, and these are readily supplied by an instrument engineer or technician. The user does not have to have programming knowledge.

Computer control can use one or more personal computers (PCs) supplied with commercially available expansion boards or interface devices, and appropriate software—the operating programs. The following are typical capabilities of microcomputer controllers, which in some cases require use of a printer:

1. Working with analog, binary, or digital inputs and outputs.
2. Data logging.
3. Strip-chart recording.
4. Plotting and storing of graphs; curve fitting.
5. Controlling the process.

With computer control, the communications between Ranks 1, 2, and 3 are generally like those of an up-to-date centralized control system.

8-3-2-2 Distributed Control

The dispersed instruments of traditional field control constituted a kind of *distributed control*. Because the instrument systems were generally functionally independent, the failure of one instrument loop did not interfere with the operation of another loop. However, the format had the accompanying disadvantages of compartmentalizing plant operations and requiring considerable manpower.

Traditional centralized control tended to counter these disadvantages but required a large number of costly tubing and wiring runs between individual local instruments and the control room. The consolidation of instruments in the control room steadily increased the information traffic that the operators handled. The effects were as follows:

1. The operators' job became easier because the operating information and important controls they needed were now close at hand. Also, their vision of plant operations was now more panoramic.
2. The large flow of information in and out, especially in a fast-changing emergency, sometimes created an overload of information for the operators. They would have to respond to one or more alarms, search for

the related system instruments, read them, evaluate the information, decide on the proper emergency action, and finally take action. The consequences of a wrong response could be serious human, physical, and financial damages, as evidenced by the Chernobyl nuclear power-plant catastrophe in 1986.

This led to the next stage of evolution: the present-day distributed control, whose development began in 1969. Distributed control is a control system that has the following dominant features:

1. Measurement, control, and communication are performed primarily by groups of modules that are distributed in function and location.
2. The control of the process by the plant operators is performed in a centralized control room.
3. Local operating control stations may be scattered over the plant.
4. If control from the control room is lost, the local operating stations continue to function.
5. If one local operating station fails, all the other local operating stations continue to operate.
6. A common communication channel, the *data highway,* runs through the plant and connects to all parts of the distributed control system (DCS). Usually, there is a second independent data highway for added reliability (see Figure 8-5).
7. A distributed control system can start small and expand as needs require and circumstances permit.

Figure 8-5 illustrates a representative DCS, which includes subsystems that are situated in several basic locations.

Function Areas by Rank

- *Plant Operation—Rank 1.*

  1. *Central Control Room.* Here is the focal point for controlling plant operations. Control rooms have evolved so that a number of PCs, with their video terminals, keyboards, and mouses, are arranged on an operator's control room console, whose length varies depending on the number of PCs.

     Consoles carry both DCS and programmable logic controllers (PLCs). Advanced system controllers, if any—for fuzzy logic, neural networks, or other—may also be located in the control room (see Chapter 5). Peripheral devices—printers, tape recorders, plotters, sequence recorders, and other specialties—may be placed where needed to accompany selected PCs. Individual printers may record either routinely or on demand.

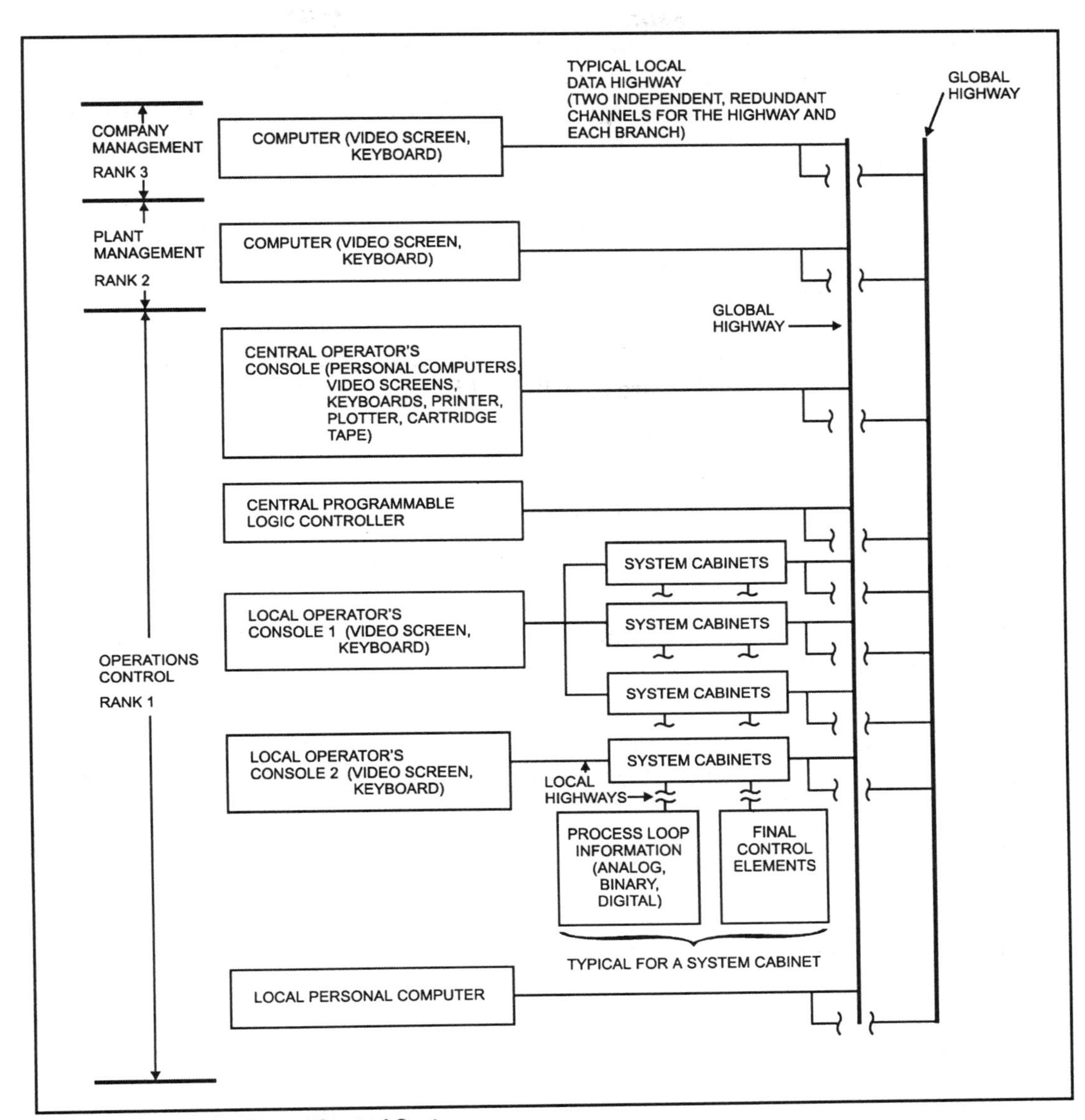

Figure 8-5. A Distributed Control System

In the control room, an operator may have access to additional information obtainable from the field or other sources. The chief operator has overall control of the process operation of the plant.

Near the control room, there is commonly a separate room for electronic devices, switchgear, and other auxiliary equipment.

2. *Local Areas.* There is no standard layout for local control instruments. A possible arrangement may include some of the following features: The system has local instruments for measurement and/or control that connect directly to local system cabinets. These cabinets may have their own PCs with video terminal, and so forth. The cabinet

information may be connected to a local console. Local control devices may respond to commands from a local cabinet or from a local console, if available. There may be channels for two-day communication with the central control room. A local PC may be tied into the DCS.

3. *Plant Laboratory.* The laboratory may have a PC connected to distributed-control highways to receive and provide the following information:

   a. Chemical analyses.

   b. Operating test data.

   c. Other process information.

   d. Historical data.

- *Plant Management Area—Rank 2.*

  This is an office in which plant planning and business are conducted. It contains computers that provide access to the information that is developed or conveyed at the central operators' console, though generally the management does not get involved with operating details. The office may be in an auxiliary building on the plant site.

- *Company Management Area—Rank 3.*

  This area is generally at the main or regional headquarters of the company. The area contains computers, and the management may communicate with the plant through distributed-control highways, radio, telephone, and other means.

8-3-2-3 Distributed Control Systems (DCSs) and Programmable Logic Controllers (PLCs)

The term *distributed control* covers two different types of control: *distributed control (DCS)* and *programmable logic control (PLC)*. Each began independently of the other and each was designed to fit different control problems. DCSs were microprocessor-based replacements for panelboard instruments, primarily for continuous control, as in refineries. PLCs were microprocessor-based replacements for wired electrical relays and mechanical timers, primarily for uses such as the manufacture of individual parts. Each type was optimized for its own needs.

The DCS controllers would handle many analog inputs and outputs, and they would be used for human-machine interfaces or operator video terminals. They used languages, such as *function blocks,* that were intended for continuous control. On the other hand, PLCs were built to perform high-speed discrete control in a factory. They used *electrical relay ladder programming*. (For a PLC example, see Section 7-4-2 and Figure 7-5.)

Over time, both DCS and PLC manufacturers recognized the changing demands for user flexibility, so each group gave ground from their formerly rigid offerings. But DCS had problems adapting to PLC, and PLC had trouble adapting to DSC; users had comparable difficulties. Many problems arose in design, configuration, start-up, operation, and maintenance. A whole spectrum of applications required both continuous and discrete controls, and user needs remained unsatisfied. The results were described as "the worst of both worlds."

In 1992, the International Electrotechnical Commission (IEC) released a specification, IEC (6)1131-3, *International Standard for Programmable Controllers.* This established the following languages: *function blocks, ladder logic, sequential function chart* (SFC), and *structured text and instruction list.* The specification states that no single process control language is best for all aspects of control:

1. For continuous control, function blocks are best.
2. For discrete control, ladder logic is best.
3. For sequencing, a graphical-sequencing language (sequential function charts) is best.
4. For programming—as for optimization routines—a Pascal-like programming language (structured text) is best.

The IEC specification also states that these languages must interact in order to provide a comprehensive process control solution. It provides for independent function-block sheets to interact with independent ladder-logic sheets and with the other languages, and also to intermix the languages on a single sheet (see Figure 8-6).

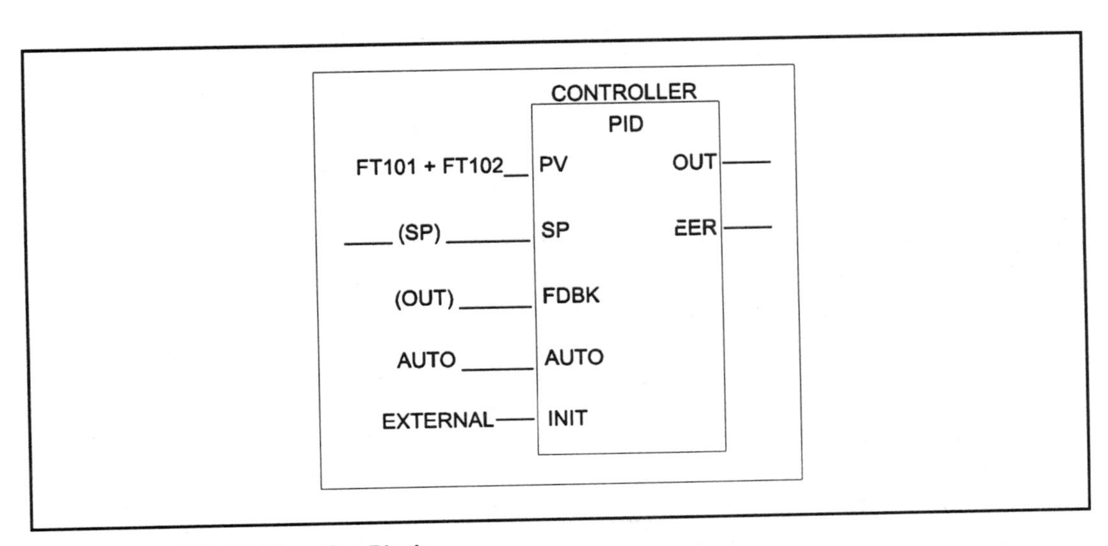

Figure 8-6. A Hybrid Function Block

8-3-2-4 Hybrid Control

Computer and process control standards have now taken hold, thus easing technical problems. Manufacturers are trending toward following the IEC lead and are creating systems that provide features and functions needed for each area of a plant and for the plant as a whole. This is accomplished by using a so-called *hybrid-control system* (HCS), which has the following features:

1. The system can be tailored to the needs of each area.
2. Where appropriate, automated start-up and shutdown operations are integral to the controller.
3. Redundant and nonredundant controls can be either physically close to the controller or remote.
4. Interaction is easy between, for example, pumps and valves that are controlled by ladder logic and by analog controls that use function-block language.
5. Maintenance people can attack a system problem by referring to the specific program that was used to configure the system controllers.
6. All language variations are consistent and can easily be read or written.
7. All electrical communications are based on standards that permit components to be commercially purchased, replaced, or upgraded.
8. The entire system has no proprietary addressing scheme.

A well-designed HCS will address the hurdles encountered through the combination of DCS and PLC systems. HCS uses a consistent system and a total-system approach to meet all diverse requirements while providing a uniform operator interface and an integrated configuration. Communications are secure at all levels.

8-3-2-5 Distributed Control Communications

A DCS requires more intercommunication between its elements than do other systems. The multiplexed data highway is the key to the whole system (see Section 3-2-5). The highway may use cables of either electrical wire, which carries electrical pulses, or of fiber optic, which carries pulses of light.

The highway ties together all the instruments and people who are concerned with the operation and management of the plant. It is the path through which all the multitude of electronic signals flow, then to be sorted and translated at their various destinations into suitable images, records, actions, reactions, and interactions. Relocated or additional instruments can be easily tied into the highway at any point. There is no need to run additional wires from these instruments to the control room, as would be required for the traditional centralized control.

The general term *data highway* covers two subcategories: *global highway,* which is a common channel that runs throughout the plant, and *local highway,* which connects a group of devices in a local area to a global highway and may handle communications within the group. Figure 8-5 shows a system with two global highways and multiple local highways.

Reliability for the system is provided by fully redundant second highways for global and local service. The dual highways may both be active or one may stand by to be switched into operation automatically if the primary highway fails.

Data highways may be miles in length, and they may carry information for thousands of control loops. But the heavier the traffic, the slower is the system performance. Heavy traffic may require one or more additional highways.

8-3-2-6 Control Video Displays

Video displays for the central console include the following possibilities:

a. This may be a diagram of a selected section of the process plant, showing process data. Or it may consist of a group of graphs for process variables from a section of the plant, showing deviations of the variables from their set points.

b. This may be an enlargement of a smaller section of the plant, showing additional data and permitting control changes by the operator. Or it may show a process change by digital readings and a graph.

c. This may be a list of tuning and alarm settings. The operator can alter these settings.

8-3-3 Distributed Control Benefits

Some of the overall benefits of distributed control are the following:

1. *For the Plant Operators—Rank 1.*
   a. More informed and timely supervision of plant operations.
   b. Quicker response to emergencies.
   c. Automatic operating reports.
2. *For the Plant Manager—Rank 2.* Easier and earlier reports concerning
   a. Abnormal events.
   b. Production rates.
   c. Consumption of materials and energy.
   d. Process efficiency.
   e. Optimization of minimum cost or maximum output.

f. Costs of production.
g. Deviations from government or customer requirements.
h. Historical data.
i. Production scheduling.
j. Inventories of raw materials and products.
k. Quality control.

3. *For the Company Manager—Rank 3.* More timely information on plant production, costs, and other factors to enter into the business plan to, in turn, strengthen market position, maximize profit, and supply necessary reports. Information may be presented in written or graphical form.

8-3-4 Local Area Networks

In order to handle the fast-growing complexity of data communications, *local area networks,* known as *LANs,* were developed. The area covered by a LAN is usually within a range of 50 kilometers (31 miles). A variety of LAN systems exist, and each is unique in operation, hardware, and capabilities. LANs offer a variety of benefits, including the following:

1. Improved staff communications.
2. User sharing of programs, data, and hardware.
3. Access to multiple databases.
4. Access to remote systems.
5. The possibility of linking machines of different vendors.
6. Possibility of access to data files of computers having different operating systems.
7. Simplified management of software licenses.
8. Use of lower-cost "diskless" workstations.
9. Reduction of cabling cost.
10. Improved security and data integrity.

Various communication standards establish the *protocols,* that is, the rules for internal and external communications within the world of computer devices. In the United States, standards have been promulgated by the Institute of Electrical and Electronic Engineers (IEEE) and worldwide by the International Standards Organization (ISO). The U. S. representative to ISO is the American National Standards Institute (ANSI). Communication models, among the many, include those of IBM Corporation, Digital Equipment Corporation, the Defense Data Network, and the Internet, Ethernet, and others. Some of these are proprietary, some are open.

8-3-5 SUPERVISORY CONTROL AND DATA ACQUISITION (SCADA)

Many industries and their branches are scattered far and wide, perhaps hundreds or thousands of kilometers from one end to the other. A central facility such as an oil field, a pipeline system, an electric transmission system, or hydroelectric plants, may find it necessary to make adjustments to equipment that is far away or difficult to access or to gather remote data from routinely. Such adjustments may be to change control set points, to operate valves or switches, to monitor alarms, to make periodic measurements, or to provide other special attention. Such situations may require the use of skilled personnel, travel, maybe helicopter access, and the expenditure of time.

SCADA (rhymes with "raid uh") is a two-way system that enables a plant operator to make the necessary remote changes or to provide the desired measurements right from his or her post in the central plant. A sample system may have the following features:

- An operator will be in a central plant of the operating company. He or she has a computer with a keyboard (or trackball or mouse). An *operator interface device* and *input/output device* (I/O) uses the keyboard for input, and output comes from a CRT (a *cathode ray tube*), also known as a VDU (a *video display unit*). The operator interfaces with the *master terminal unit* (MTU), which is the system controller. This controller can be programmed to repeat instructions periodically.

- The single MTU is connected to a number of *modems*, each of which (a) translates digital pulses from the computer into analog signals for each field device that is being supervised by the MTU and (b) translates analog signals from the field into digital pulses that convey information to the MTU. Each field modem connects to a *remote terminal unit* (RTU).

- A SCADA system may contain a number of RTUs, ranging from one to hundreds. Each single RTU must be able to understand the following:

  a. A message has been directed to that RTU; the message must be decoded; the message must be acted upon; it must be decided whether a response to the message is required; a suitable response must be made, if necessary; when the work is finished, the action should shut down; and the RTU should be ready to act if a new message arrives; and

  b. performing these actions might require, for example, checking the existing position of field equipment; comparing the existing position to the required position; deciding whether any action should be taken; sending an electrical signal to a field device to order it to change state, if necessary; checking switches to ensure that the order was obeyed;

and sending a confirmation message to the MTU that the required new condition was reached.

- There are two common media of LAN communication: land line, in the form of (a) fiber-optic cable or electrical cable or (b) radio. Large systems may use a combination of radio and telephone lines. Usually the MTU will have auxiliary devices, such as printers and backup memories, that are considered to be accessory to the MTU.

- In many applications, the MTU sends accounting information to other computers or management information to other systems. These connections may be direct or in the form of LAN drops. In a few cases, the MTU receives information from other computers. This is especially true of newer systems in which application programs, operating on other computers and connected to the SCADA computer, provide a form of supervisory control over SCADA.

8-3-6 An Assessment

The advent of distributed control has caused important changes in the design and use of control systems, especially for large plants. But the resulting improvements in performance are a matter of degree, not of kind. The fanciest hardware in the world used with an inferior control strategy may bedazzle us with its technology, but it will be outshone in process results by humdrum, old-hat hardware that uses the right control strategy. There is indeed a place for the latest super-duper electronic wonders, but there is no substitute for good control thinking, which is the bedrock of process control. A general policy should be to apply the simplest, technically satisfactory solution when solving a given problem.

On the other hand, assuming equal control strategies are employed for a traditional hardware system and a DCS, the latter offers the following important benefits:

1. Improved control of the process by the plant operator.
2. Better communication between the company's Ranks 1, 2, and 3.
3. Greater flexibility with easy changes of control plan.
4. Less hardware because of the electronic concentration of functions.
5. Lower costs for installed signal lines and hardware.
6. More reliability through the extensive use of redundancy and the semiautonomy of local instrument loops.
7. Less control room crowding, with a reduced need for utility services to provide comfort.

Traditional control systems may be made up of instruments from many different manufacturers and are frequently purchased that way. These instruments have a high degree of compatibility and interchangeability. It is up to the traditional system designer to ensure the compatibility of several instruments in a loop, for example, and usually this is not a problem.

The situation is different for distributed control, for which all the devices that are or will be directly linked to the data highway should come from one manufacturer and should be bought as a system in order to provide signal compatibility. (See the following ISA Standards: S50.1, *Compatibility of Analog Signals for Electronic Industrial Process Instruments;* S50.02, *Fieldbus Standard for Use in Industrial Control Systems, Part 2: Physical Layer Specification and Service Definition;* S72.01, *PROWAY-LAN Industrial Data Highway;* S72.02, *Manufacturing Message Specification: Companion Standard for Process Control.)*

9

External Influences on Instrumentation

Instruments and their interconnecting wiring and piping, including tubing, must be able to withstand, to some acceptable degree, the effects of the process conditions that the devices may be directly exposed to. The exposure may be caused by inserting instruments into the process fluids or by piping the fluids to the instruments. This closeness may subject the devices to severe conditions—extremes of temperature, pressure, velocity, corrosiveness, and other factors related to the process fluids. All these things must be taken into account when choosing instruments and their methods of use.

Furthermore, there are environmental or ambient factors (*ambient* means "surrounding") that must also be taken into account. These factors may be unrelated to the process or related only indirectly; for example, a pipeline may give off much heat and make the nearby surroundings hot.

All these external factors may affect the materials of the instruments or their accuracy, or both. Similarly, the factors affect the materials of wiring and piping. Some of the most important effects on instruments' materials and accuracy are discussed in the next two sections.

9-1 Effects on Materials

9-1-1 From Temperature

The plant environments in which instruments may be placed range from an arctic -50°F, or lower, to a tropical 140°F, or higher (-46°C to 60°C). The minimum temperature of an instrument may be as low as the lowest outdoor air temperature. However, the maximum temperature may exceed the highest outdoor air temperature because of heat received directly from solar radiation or nearby heat sources.

ISA Standard S71.01, *Environmental Conditions for Process Measurement and Control Systems: Temperature and Humidity,* establishes classifications for location and severity of environmental temperature and humidity. Protective enclosures for heating or cooling may be required or, when feasible, the instruments may be relocated. Electrical parts, such as transistor circuits, computer chips, and operating coils for solenoid actuators, are sensitive to high temperature, which can shorten their effective life. The manufacturers' instructions for household electronic goods, such as a radio or a computer, often warn the buyer against keeping the goods in a hot location. Manufacturers specify different ambient temperature limits for their instruments. For example, the limits may be as follows:

1. For operation: +40°F to +122°F (+5°C to +50°C).
2. For storage: -40°F to +150°F (-40°C to +65°C).

For pneumatic field instruments, a manufacturer specifies an environmental temperature range of -40°F to +180°F (-40°C to 82°C).

Metals lose strength and soften as they get hotter. At some elevated temperature, their crystal structure changes, resulting in physical changes that persist after the temperature falls. The metals then require heat-treating to restore the original properties, and such treating is usually not feasible.

Metals that are chilled to very low temperatures become prone to fracture from brittleness. For example, at temperatures below -20°F (-29°C) stainless steel is often used instead of ordinary carbon steel because it is much less subject to brittle fracture. The brittleness does not persist when the metals are warmed. Plastic parts, such as gaskets, deteriorate faster and tend to become irreversibly brittle as a result of chemical change at high temperatures. They become reversibly brittle from simple hardening at very low temperatures.

The passage of time causes many materials to deteriorate. This aging effect occurs more rapidly at high temperature than at low temperature.

9-1-2 From Humidity

Humidity is the content of water vapor in the air and is one of the environmental use factors that instrument manufacturers specify. When liquid water evaporates, it goes into the form of vapor, which is always invisible. (A water fog or mist, loosely called "steam," over a pond or a pot of boiling water is actually formed by minute particles of liquid water.)

The amount of water vapor that is present in air is the *absolute humidity,* expressed in pounds of vapor per cubic foot of air or equivalent units. The absolute humidity can vary from zero to some value that depends on the air temperature; the higher the temperature, the more moisture can be retained in the air. That is why the moisture in a warm room condenses on a cold outside window. At a given temperature, air is said to be saturated when it contains the maximum amount of water vapor that can be retained at that temperature; it cannot dry up a water puddle.

Atmospheric air is usually not saturated, so it contains less than the maximum quantity of water vapor for the existing temperature. *Relative humidity* (RH) is the ratio of the actual amount of vapor present (per cubic foot of air) to the greater amount of vapor that would be present if the air were saturated. Relative humidity is expressed as a percentage and can vary from zero, which is for so-called bone-dry air, to 100%, which is for saturated air. Relative humidity is normally what is intended when a weather report speaks of "humidity." The relative humidity, not the absolute humidity, is what makes us feel comfortable or uncomfortable.

Electronic instrument manufacturers warrant their instruments for service typically for a humidity range of 5% to 95% RH when there is no condensation. A humid atmosphere inside an instrument may cause moisture to condense on the instrument components. This may lead to instrument damage from an electrical fault (a ground or short circuit). However, an instrument containing electrical parts that are energized in normal operation receives heat from those parts. The heat lowers the relative humidity of the inside air even though it does not lower the absolute humidity. This reduces the possibility of condensation.

A humid atmosphere also promotes corrosion and the deterioration of susceptible materials. Some instruments that are intended for a hot humid climate are *tropicalized,* that is, designed and treated to protect them against that environment.

9-1-3 From Explosion

Certain vapors or gases (such as benzene or methane) or dust from solids (such as coal, grain, or wood) or fibers can burn or explode if they are mixed with air and ignited. Such mixtures are called *flammable* or *combustible* although they have, illogically, also been called *inflammable.*

For a flammable vapor or gas to burn with air—explosion is an extremely rapid burning—there must be a combustible mixture. Only then can there be the chemical reaction that constitutes burning. The concentration of each flammable substance must be within a specific range in order to sustain burning. The limits of the range are only approximate because they depend on the temperature, pressure, humidity, and other components of the mixture. A few examples of the lower and upper limits of flammability are given in Table 9-1.

Table 9-1. Representative Flammability Limits in Air

Flammable Substance	Flammability Limits, Percentage by Volume	
	Lower	Upper
Acetylene	2.5	80.0
Carbon monoxide	12.5	74.2
Coal dust	4.8	33.5
Natural gas	4.8	13.5
Propane	2.2	9.5

For an explosion to begin in a mixture, all of the following must exist simultaneously:

1. A flammable substance must be present.
2. Air or other oxidizing substance must be present.
3. Mixture must be within the flammable range.
4. There must be a starter source of ignition, which may be either an electric spark or a high-temperature spot.

One of the advantages of pneumatic instruments in potentially explosive areas is that the instruments do not contain an ignition source unless they contain electrical components, such as switches. Certain instruments, such as chemical analyzers, are unavoidably electrical in design. But even where avoidable, the potential hazard from explosion does not necessarily rule out the use of electrical instruments. The electrical instruments can be put inside explosion-proof enclosures or be otherwise made safe.

Methods of protection against possible hazards in a plant that handles flammable substances include the following six techniques:

- *Use properly designed and properly maintained process equipment, operate the equipment properly, and maintain good housekeeping.* The idea here is to avoid even the beginning of trouble.

- *Use continuous automatic analyzers.* If such an analyzer is available to detect a hazardous substance in the atmosphere, it can be used to give an early warning of the presence of the substance in case of a leak. The operator can then take protective action before the lower flammable limit is reached. Obviously, to rely on the analyzer, it must be kept in good working order.

- *Use wholly pneumatic instruments.* This eliminates the possibility of an instrument creating an electric spark in the hazardous area. Wholly pneumatic switches and alarms, both audible and visible, are commercially available.

- *Use explosion-proof housings.* (In Europe, these housings are called flameproof.) Such a housing for an instrument or a component does not necessarily prevent an electric spark and a resulting explosion from occurring. But if a mixture is ignited inside the housing, the explosion is confined within the housing by two means: (a) the housing is structurally strong enough to withstand the internal explosion without being damaged, and (b) it has a closure joint that is tight enough to permit only the slow release of the resulting combustion gases.

- The National Electrical Code (NEC) has established categories for *hazardous locations*, which are defined as locations in which combustible vapor, gas, dust, or fibers may be present in explosive proportions. The classifications are as follows:

    1. Class I, II, or III, based on the physical form of the combustible material: gas or vapor, dust, or fiber.
    2. Division 1 or 2, based on whether explosive concentrations are present or apt to be present or whether they are contained but may be present in an abnormal situation.
    3. Groups A through G, based on the specific kind of hazardous substance that may be encountered.

Other related information is also available.

- *Prevent the explosive mixture from making contact with the ignition source.* This may be done for flammable vapors and gases in one of the following ways:

    1. The manufacturer encloses a switch or other electrical part in a hermetically sealed container.
    2. A *purge*, which is a small flow of clean air or inert gas such as nitrogen, is used to sweep out the hazardous substance. The purge gas pressure must be maintained high enough to prevent backflow, and adequate safeguards must be provided to guard against failure of the gas purge. Depending on the situation, other requirements may exist.

- *Use intrinsically safe instrument systems.* This method is based on the fact that a combustible mixture must be heated to its ignition temperature before the mixture can explode. A spark can be so weak that it releases insufficient heat to sustain combustion, so nothing really happens. On the other hand, a strong spark releases enough heat to burn enough combustibles to initiate a self-sustaining reaction, and the explosion follows. To be intrinsically safe, the potential energy release must be so low as to be safe under all normal conditions and any conceivable combination of abnormal circumstances or faults.

9-1-4 From Environment-borne Contaminants

There are several general types of contaminants that may affect instruments through chemical, mechanical, thermal, or electrical effects. They are described in the following two sections.

9-1-4-1 Airborne

1. *Liquids.* Liquids may be carried to instruments and their appurtenances by rain, snow, or plant washdown.

2. *Vapors.* Vapors of water, solvents, lubricants, and other chemicals can migrate through the atmosphere. They may condense, ending up as droplets or puddles.

3. *Aerosols.* These are suspensions of minute droplets of liquid that form a mist containing salts, as from brackish water (i.e., containing salt) or seawater. The aerosols may be created by water-cooling towers, by the action of waves or wind, or by emissions from stacks.

4. *Solids.* Dust is a universal contaminant of environmental air and other gases. The dust may originate from a large variety of sources including textile fibers, furnace fly ash, plastic powders, tobacco smoke, and road dust. Some manufacturing processes, such as those for certain electronic parts, require almost perfectly dust-free air. The work is carried on in so-called *clean rooms*, which require a very special design and extreme and continuing precautions to maintain cleanliness.

5. *Gases.* Chemical pollutants originate from transportation vehicles, industrial plants, trees and other flora, and that innocent place—the home, which releases chlorine compounds from bleach, organic cleaning compounds, combustion gases, ammonia, gaseous by-products from burning tobacco, and ethyl alcohol from alcoholic beverages. Our source of life, the sun, does its bit through its ultraviolet rays, which convert some atmospheric oxygen to ozone, a powerful chemical reagent. The ultraviolet light also catalyzes (promotes) the interactions of chemicals and creates still other chemicals. So far as our atmosphere is concerned, we can say there is a real chemical jungle out there.

9-1-4-2 Biological

Biological factors include the following:

1. Birds and insects that build nests in and plug vent lines.

2. Rodents and insects that chew away electrical insulation.

3. Dead creatures, fungi, and mold. (Shortly after World War II, the electric computers of the day used electromechanical relays and vacuum tubes. When a computer failed, the people in charge found that the trouble was caused by the body of a dead insect, which prevented the electric contacts of a relay from operating. The system had a bug in it. They removed the carcass, thus "debugging" the system, which worked again. The words *bug* and *debug*, relating to a system defect and its correction, have become universalized and are now part of the English language.)

4. Marine life that builds deposits on sensing probes in pipes and on heat-transfer surfaces.

The pollutants, some by themselves and others by making use of the moisture in the air, can affect instrumentation and may lead to the failure of part or all of an instrument system. The effects include the following types:

1. *Chemical.* Weakening or pitting a smooth part or creating pinholes.
2. *Mechanical.* Accelerating the wear of moving parts; increasing friction, thus interfering with sensing.
3. *Thermal.* Fouling the heat-transfer surfaces, leading to reduced heat-transfer efficiency and sometimes plugging the heat-transfer tubes.
4. *Electrical.* Short-circuiting, grounding, or insulating electrical contacts.

ISA Standard S71.04, *Environmental Conditions for Process Measurement and Control Systems: Airborne Contaminants*, establishes classifications for the type and severity of airborne contamination.

To protect instruments against harmful atmospheric effects, the National Electrical Manufacturers Association (NEMA) Standards Publication No. 1S 1.1 has established classifications for enclosures for electrical devices, including instruments. The NEMA classifications are used for nonelectrical instruments also. The classifications begin with Type 1, which is called *general purpose* and is intended for use indoors in a normally "clean" atmosphere. Classification Types 2 through 13 cover the gamut of enclosures that are dust-tight, drip-proof, rain-tight, rainproof, sleet resistant, sleet-proof, watertight, submersible, corrosion resistant, oil immersed, oil-tight, or for flammable service. Enclosures satisfying these various classifications are available commercially. In some cases, a roofed open shelter may suffice.

9-1-5 From Radioactivity

Radioactivity is the emission of nuclear particles. It comes from radioactive substances on Earth and from the sun, which is a huge nuclear furnace. It exists virtually everywhere, including the home, as weak background radiation. In normal everyday doses, it may be harmful to people and things over the long term but the extent of the harm, if any, is still under scientific debate.

Radioactivity can become a problem in a plant that generates nuclear power or that processes nuclear substances for fuel, medical purposes, or weapons. Small quantities of nuclear substances are also used occasionally in ordinary, nonnuclear process plants to measure the density of materials or the level in a vessel as well as to check the quality of metal welds. In such uses, the local intensity of nuclear radiation may be high and damaging. There is no scientific argument about this. Radiation warning signs should be placed where this hazard exists.

The effect radioactivity has on organic materials—plastics, oils, and others—depends on the material, but generally it reduces their strength, increases their hardness and brittleness, causes swelling and discoloration, makes them gummy, reduces their lubricity, and induces radioactivity in them. Metals may suffer embrittlement, change of strength, or transmutation (so that a homogeneous metal ends up as an alloy), and the metals may become radioactive.

Instruments that will be used in a nuclear environment should use only materials, so far as possible, that can withstand the expected radiation over an acceptable period of time. For example, steel or stainless steel is usually acceptable, while aluminum, mercury, and zinc are not acceptable. Polypropylene plastic is acceptable; Teflon is not acceptable.

If a suitable long-life material is not available for an instrument part, an appropriate replacement schedule should be followed to ensure proper functioning of the part.

9-1-6 From Shaking

Instruments are mounted on supports: pedestals, boards, piping, walls, machinery, and so on. Machines and motors vibrate and cause the structural members of the building to vibrate. This causes everything attached to the building to vibrate. A pipe may vibrate from fluid flow or from an unstable control loop whose control valve *hunts*, that is, oscillates; this causes everything attached to the pipe to vibrate. Whatever the source of vibration, the instruments often vibrate to some degree.

Plant vibration mechanically wears the moving parts of instruments. It also causes *fatigue*, which results from repeated or prolonged stress and may cause a part to crack or break. A familiar example of failure from fatigue is the breaking of a paper clip by bending it repeatedly. In most cases, instrument vibration is small enough so that fatigue failure does not occur.

When an earthquake occurs, it creates a shaking effect that differs from the normal plant vibration. The earthquake swings are much wider and usually slower. This shaking does not cause fatigue inasmuch as it does not last very long, but it may be so severe that it causes permanent deformation or breakage of the parts or supports of the instruments.

To avoid instrument damage from the shaking caused by normal vibrations and earthquakes, many instruments are available that withstand severe movements and shock without failing. Shock absorbers are sometimes used in these instruments, for example. Instrument tubing, electrical cables, and their supporting trays and base supports must also be able to withstand plant vibration or shaking.

The ability of instrument systems to keep functioning during and after earthquakes is very important for the systems of nuclear power plants that are related to nuclear safety. The designers of these plants and the manufacturers of the associated instruments take great pains to provide overall system integrity.

The same kind of care may be appropriate for important systems in other kinds of plants.

9-1-7 From Lightning

Outdoor instruments may require protection against the damage caused by lightning strikes. All electrical instruments, especially those outdoors, may require protection to prevent internal damage as a result of voltage surges caused by lightning.

9-2 Effects on Accuracy

9-2-1 From Temperature

Ambient temperature may affect the properties of mechanical parts, such as a bellows, or electronic parts. This may result in a loss of instrument accuracy as the temperature changes from its design value. Two typical manufacturers' statements of the effect that ambient temperature changes have on the accuracy of their instruments are as follows:

- ±0.28% of reading per 50°F (28°C) change, based on a reference temperature of 77°F (25°C).
- ±0.25% of span per 50°F (28°C) change.

Sensing lines filled with liquid are frequently used to measure flow or level for pressure or differential-pressure measurements. These lines are subject to ambient temperature changes from day to night as well as from changes in the weather. The changes heat or cool the liquid in the lines, thereby changing the liquid density and the resulting pressure that the instruments sense. Even if the instrument itself is protected and kept at a constant temperature, the effect on the external sensing line causes measuring errors that may be significant unless precautions are taken.

Changes in ambient temperature raise or lower the temperature of the capillary and pressure sensor in a filled system, which may be used to measure process temperature, pressure, or level. Since a filled system is really a pressure system (as explained in Section 3-7-2-4 on filled thermal systems). The result is to change the measurement. Ambient temperature effects can be nullified if the instrument is furnished with automatic temperature compensation.

All the temperature-induced errors are reversible and disappear when the ambient temperature returns to its design value.

9-2-2 From Humidity

Ambient humidity (see Section 9-1-2) may cause electrical leaks that do not necessarily stop electrical instruments from functioning but that are hidden causes of instrument errors.

9-2-3 From Pressure

Ambient pressure affects the accuracy of gage-pressure instruments (see Section 3-4-2).

9-2-4 From Environment-borne Contaminants

These contaminants were discussed in Section 9-1-4. Ways that they may affect instrument accuracy include causing the leakage of electrical signals, increasing mechanical friction, interfering with electrical switches, plugging vents, and slowing the dynamic response of temperature sensors because of the buildup of external films.

9-2-5 From Lightning and Other Electrical Phenomena

Lightning may cause transient voltage surges or the momentary loss of voltage in electrical circuits. These effects may cause a loss of circuit memory or a change of information in electronic storage. Protecting against these effects may include using surge suppressors and backup power supplies.

Electronic systems may also be affected by nearby electrical noise—so-called *radio-frequency interference* (rfi) and *electromagnetic interference* (emi)—which may be produced by electrical power cables, power-line fluctuations as machines swing on and off, wireless and cellular phones, and portable tools. Protection against these effects may be provided by noise filters and proper electrical wiring practices, including shielding, grounding, and barriers against or separation from noise sources.

9-2-6 From Gravity

Certain measurements—of pressure and other variables—may conceivably be in error by as much as 0.5 percent of the true value because of gravity differences. This is rare but could occur if the sensing instrument is used in a geographic location that is very far from the place where it was calibrated. When calibrating an instrument by means of a standard weight, correction should be made for the effect of gravity on the weight, if necessary. (See the discussion of force and weight in Section 3-4.)

9-3 Power Supply Requirements

Power supplies—electrical, pneumatic, or hydraulic—are needed to enable most instruments to function. For the instruments to function properly, the power supplies have to meet the manufacturer's specifications. These are described in the next three sections.

9-3-1 Electrical Power Supplies

Representative requirements for voltage and frequency are as follows:

1. 120 V, +10% to -15%, ac and 50/60 ± 2 Hz (*hertz*, formerly *cycles per second*).
2. 125 V, +15 to -20 V, dc.
3. 24 ± 2 V, dc.

9-3-2 Pneumatic Power Supplies

A representative manufacturer's specification for a pneumatic instrument calls for the air supply pressure to be as follows:

1. Normal: 5 psi above the maximum pressure of the instrument output.
2. Minimum: 3 psi above the maximum pressure of the instrument output.
3. Maximum: 50 psig.

The purity of the instrument air may be important, especially if an instrument has small air passages. The air supply comes from a compressor whose air discharge is usually relatively humid and often carries droplets of compressor oil and dust particles. This air is called *plant air.* The droplets and dust may coalesce or agglomerate. The condition is worsened when condensation or freezing occurs in the air system. The result may be to clog the air passages of pneumatic instruments so that the instruments operate poorly or not at all.

The normal practice for pneumatic instrumentation is to provide an air purification system. This is required even in most electronically controlled plants, which usually require pneumatic control valves and some local pneumatic instruments. The product of the purification system is dry and clean *instrument air.* For instrument air, the word *dry* means having a relative humidity so low as to prevent condensation of the moisture in the air at the lowest plant temperature to which the air may be exposed.

The purification system is supplied with plant air, which is passed through filters to mechanically remove liquid and solid particles and driers to remove water and other vapors. Some driers dry by using either refrigeration or expendable chemical agents, such as lime. Another type of drier uses desiccant (drying) pellets of alumina or silica gel that retain the vapors from the air passing through the drier. This type of desiccant can be regenerated by heat and used repeatedly.

(We sometimes encounter small bags of desiccant when we open a new package containing an expensive camera or pills that have to be kept dry.)

ISA Standard S7.0.01, *Quality Standard for Instrument Air,* establishes allowable limits for moisture, particle size, oil concentration, and other contaminants. The American Nuclear Society (ANS) Standard 59.3, *Safety-Related Control Air Systems,* discusses instrument-air systems for use in nuclear power plants. However, this standard may be helpful also for nonnuclear plants, especially for those process systems that have a particular need for reliability in their pneumatic instrument systems.

There are situations in which air substitutes may be used. An instrument far from a plant air supply may use nitrogen, which is a chemically inert gas supplied in portable steel bottles with pressure regulators. Even pipelines carrying process gases that are dry and noncorrosive are sometimes used to power local pneumatic instruments.

Plant air is sometimes used for pneumatic devices such as on-off piston actuators. These actuators are simple machines that may not have small passages that can become clogged. But plant air is not customarily used for such applications unless the point of use is remote, instrument air is not available, and plant air is available.

9-3-3 Hydraulic Power Supplies

A hydraulic instrument system uses a hydraulic liquid that has to be kept clean and at the operating pressure specified by the supplier of the hydraulic system.

10

Incidental Hardware

The previous chapters described the hardware scenario in which instrumentation systems function. The instruments are the lead actors in this production, whose purpose is to show how a process plant can be controlled. The roles of supporting cast and bit actors are played by other hardware that is important in making the production a hit and bringing the audience to its feet cheering. Prominent among the many secondary members of the cast are the devices described in the next seven sections.

10-1 Temperature Wells

Temperature wells, commonly called *thermowells* (in Great Britain, *pockets*) are metal pieces, as shown in Figure 10-1. They are customarily supplied as accessories for insertion-type temperature sensors. The well is installed projecting into a process pipeline or vessel, and the sensor is inserted into the well. The well is made from a corrosion-resistant material and is strong enough to resist mechanical damage and to stand up under the process pressure and temperature. The purposes of the well are as follows:

1. To protect the temperature sensor against corrosion, breakage from vibration caused by fluid flow, and damage from erosion.
2. To enable the sensor to be replaced without requiring that the plant operation be stopped.
3. To permit the future insertion of a sensor for either temporary test measurements or permanent use.

Thermowells have thermal resistance, which is undesirable because it slows the response of the sensor to temperature changes.

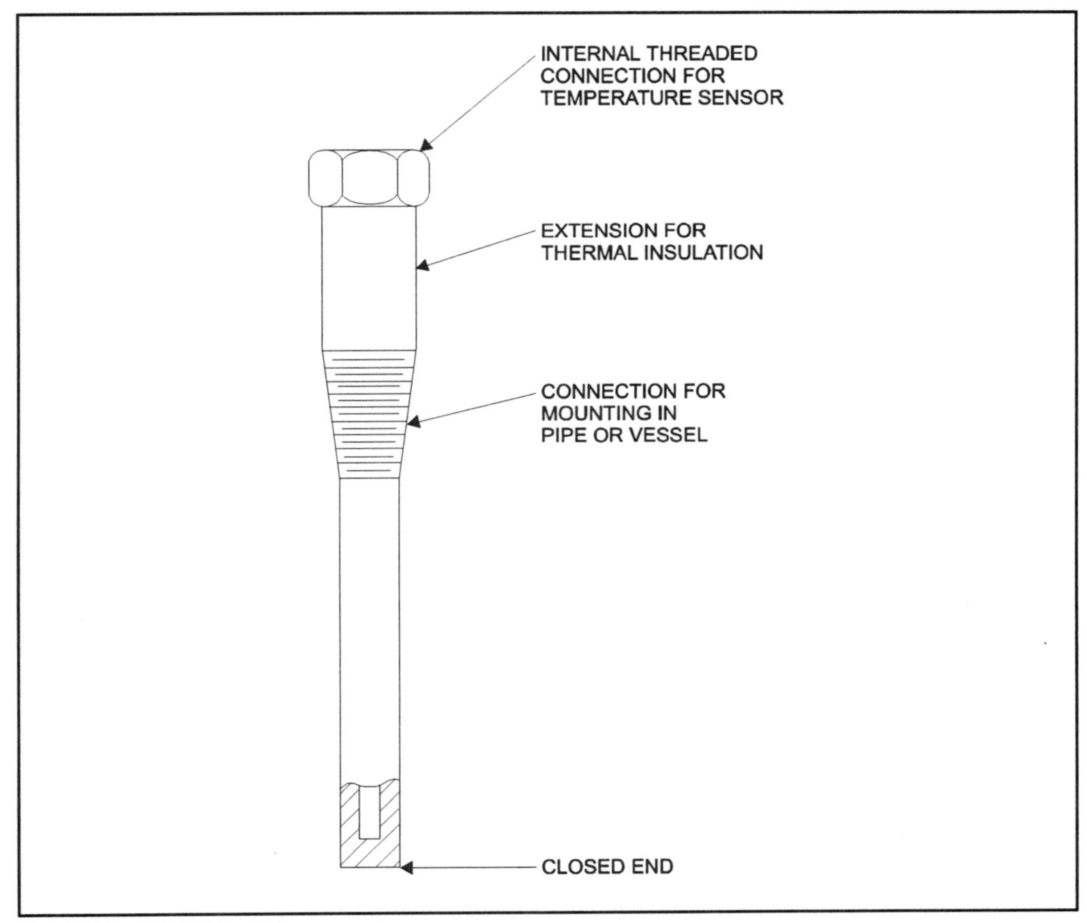

Figure 10-1. A Typical Thermowell

10-2 Isolation Valves

With few exceptions, every sensing line is provided with a *root valve*, which is an isolation or shutoff valve close to the process line or vessel. Generally, each sensing line branch running to a separate instrument has a shutoff valve placed near the instrument. This valve is known as an *instrument valve*. Each sensing line branch may also have a valve to bleed air or other gas from the branch if it carries a liquid or to bleed liquid if it carries a gas. The purpose of this valve is to keep each line or branch full of the intended fluid. A valve may also be provided for connecting a calibration or test instrument. Figure 10-2 shows a typical sensing line for a pressure instrument.

Similarly, there are shutoff valves for pneumatic or hydraulic signal lines and power supplies. Analogously, there may be isolation switches for electrical signal lines, and circuit breakers or switches for power supplies. Electrical instruments may have connections such as screws or jacks (a form of electrical socket) so they can be used temporarily with calibration or test instruments.

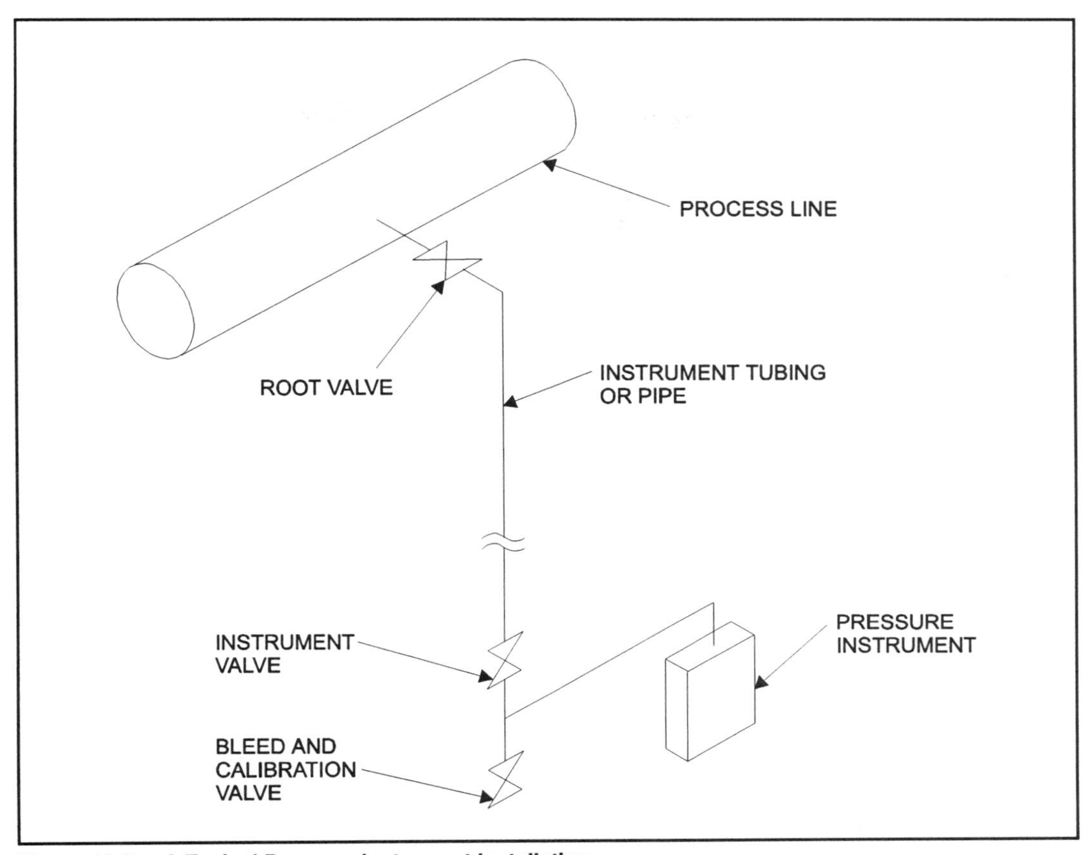

Figure 10-2. A Typical Pressure Instrument Installation

10-3 WEATHERIZING DEVICES

Many process liquids freeze or precipitate solids in cold weather. For example, sodium chloride (common salt) begins to settle out from a 36.6% solution with water unless the temperature is kept above 122°F (50°C). Some liquids boil in hot weather. These effects of extreme temperatures may present a problem for sensing lines, especially if they are stagnant, or for the instruments themselves. To cover these devices with thermal insulation is not an answer because insulation does not supply or remove heat; it only slows the cooling of hot liquids or the warming of cold liquids.

It is common practice to provide a heating system with insulation to protect susceptible individual lines and instruments against the outdoor cold. Such protection is called *winterizing*. Methods for doing this are the following:

1. *Electrical Tracing*. Electrical heating elements are run alongside and in close contact with the sensing lines and instruments.
2. *Electrical Heating*. Metallic sensing lines themselves are made to act as heating elements by passing electricity through their walls.

3. *Steam Tracing*. Small tubing is laid in contact with the lines and instruments to be heated, and the tubing carries heating steam.
4. *Enclosures*. Special enclosures are very often used for the instruments. The enclosures are heated by either electricity or steam.

Each kind of protective system needs some auxiliary instrumentation for regulating the heat input to the process instrumentation that is to be protected. Some lines and instruments may contain a volatile liquid that boils on a hot day. Such cases require a means for cooling: *summerizing*. The simplest practice is to change to a sensing technique that does not require that the sensing lines contain the volatile liquid.

10-4 Instrument Enclosures

Section 9-1-4 discussed a NEMA Standard 1S 1.1 for enclosures to protect all types of instruments against harmful atmospheric effects. Refer to that publication for a description of the thirteen enclosure classification types.

10-5 Power Supplies

To determine the amount of air flow required to operate pneumatic instruments, including control valves, it is necessary to count the number and types of pneumatic devices and connecting metal tubing the plant will use. Similarly, the amount of electrical power required to operate plant instruments is based on a count of the various types of instruments. For instrument steam tracing, if needed, the steam consumption is estimated according to the count and types of instruments that may require thermal protection during a cold environment.

Instrument air is usually distributed at a pressure of nominally 100 psig, which is lowered at each instrument as required, for example, to 20 or 35 psig. The pressure reduction is usually performed by a 1/4-inch adjustable reducing valve that has a filter, drain connection, outlet pressure gage, and overpressure protection. This valve is termed an *air set* or *air filter regulator* and is procured as an accessory for each instrument or as a bulk commodity.

A hydraulic control system may be part of a machinery package or it may be procured separately.

Determining the incidental needs of the several utilities just mentioned may be performed by the instrument department, the piping department, the electrical department, the mechanical department, or an overall construction department, depending on the size and structure of the engineering company or manufacturing company that is involved in constructing the instrumentation system.

10-6 Instrument Supports

Supports for instruments obviously have to be strong enough to hold the dead weight of the instruments and their immediate accessories, including a portion of adjacent connecting piping, tubing, and wiring. The support design may also need to allow for the weight of careless construction or maintenance workers who step on the supports.

In addition, the effects of dynamic loads on the instrument supports should be considered, as discussed in Section 9-1-6. The supports are usually designed by civil engineers.

10-7 Instrument Boards

Instrument boards come in different sizes and styles, which are summarized as follows:

1. *Consoles*, which were discussed in Section 8-3.
2. *Instrument racks*, which are open steel frameworks on which instrumentation is mounted.
3. *Cubicles*, which are large enclosures.
4. *Panels*. The term panel is frequently used ambiguously. ISA standard S5.1 stated that *panel* and *board* are synonymous, and thus *panel* is often used in the sense of *board*. However, both "boards" and "panels" have component sections that are called "panels." For clarity, this book refers to an entire structure as a *board* and its sections as *panels*.
5. *Benchboards*, which are vertical boards that have a sloping, desk-height section.

11

Design and Selection of Instruments and Systems: Engineering Considerations

We have seen how process instruments are used, singly and in combination. But before deciding on a particular kind of instrument or loop for measuring or controlling, there are engineering factors, described in this chapter, that should be taken into consideration. There are also managerial factors, which will be discussed in Chapter 12.

11-1 Loop Accuracy

In Section 3-1-5-1, we took up the subject of the accuracy of a single instrument. (We should remember that when the word *accuracy* is used for an instrument or loop, its meaning is "error," "inaccuracy," or "uncertainty." For clarity, therefore, the word *error* is used instead of accuracy in the following discussion.) What is the overall error of a measurement that is made through a series of instruments—for example, the instruments LE-1, LT-1, LY-1C, and LC-1, as shown in Figure 3-8?

If the loop exists physically, the way to determine its actual error is by testing it. This is done by giving the first element of the loop an input and seeing what comes out of the last element. However, it may be necessary to predict a loop error. Suppose the errors in this measuring loop are as follows:

Sensor	LE-1	±0.5% of span
Transmitter	LT-1	±0.25% of span
Signal converter	LY-1C	±0.75% of span
Receiving element of	LC-1	±0.4% of span

Assume that these four errors are those claimed by the instrument manufacturers. If the environmental conditions are not as specified by a manufacturer, then additional errors are introduced, as discussed in Section 9-2, and the result may become worse.

The overall error of the loop can be determined by one of the methods described in the next three sections.

11-1-1 Algebraic Error

The simplest way to determine the overall measurement error is by adding the absolute value of the four individual errors, which means disregarding the symbols for the addition. The result is the algebraic error, which in the preceding example is 1.9% of span.

This method is fast and easy. It is also very safe because it contains two implicit assumptions about the four errors:

> Assumption 1: The errors are all positive or all negative, not mixed.
>
> Assumption 2: Each error is at an extreme of the manufacturer's guaranteed error band.

The law of chance for random errors does not usually work this way, and it is unlikely that the worst-case assumptions will hold true in a given case in real life.

If an overall error of ±1.9% of span is acceptable for the service, then the matter can rest. However, if the service requires that the overall error not exceed ±1.1% of span, then the result is not acceptable. A more accurate error estimate is required. One alternative, if the required measurement range covers only a portion of the range for one or more of the instruments, is to calibrate the instruments over just that portion instead of over the full instrument range. This reduces the overall error. Some environmental errors may be reduced by relocating the problem instrument or instruments or by using some type of compensation.

11-1-2 Probable Error

The most widely used method for determining the overall error gives the *probable error*, not the extreme and unlikely algebraic error. The probable error is the error that is just as likely to be exceeded as not to be exceeded. This is calculated by taking the square root of the sum of the squares of the individual errors, which gives a result of ±1.0% of span. This is acceptable if the requirement is that the error not be greater than the allowable maximum, ±1.1% of span.

In this example, all the individual errors were given equal weight, that is, all were treated the same way in the calculation. In calculations of probable error for some instrument loops—frequently for flow measurement—some of the individual errors are given more weight or less weight than the other errors receive. The need for doing this depends on the mathematics of the measurement equation.

11-1-3 Statistical Error and Probabilistic Error

In a given case, the probable error may be unacceptably large, and the algebraic error is even larger. Or the very concept of probable error may be unacceptable because, by definition, the concept acknowledges a one-out-of-two chance that the actual error will be worse than the probable error. If the measuring system cannot be improved to reduce the error, then more refined methods may be used to determine the overall error.

Such methods are the *statistical method* and the *probabilistic method,* which are quite sophisticated and may involve special testing and fallible human judgment. These methods are used rarely and only when dealing with especially important cases. They yield what may be the most realistic estimates of error, but these will not necessarily be any more acceptable than the others were.

11-2 RELIABILITY

11-2-1 WHAT IS RELIABILITY?

Reliability is the probability that a product will provide a *specified performance* when operating under *specified operating conditions* for a *specified length of time.* In other words, the reliability of an instrument or instrument system is the consistency with which it measures or controls as it is supposed to under the right conditions and according to its program and adjustments.

Two important terms concerning reliability are the following:

1. *MTBF,* which is the *Mean Time between Failures* of a given type of instrument or system, as determined by test, experience, or both. *Mean time* means the *average time.* A large MTBF is good and depends on (a) the equipment manufacturer using high-quality materials, conservative design, and care in manufacture and (b) the operating plant applying the equipment to the kind of service for which it was designed and performing routine maintenance as recommended.
2. *MTTR,* which is the *Mean Time to Repair a Failure,* as determined by experience. A small MTTR is good and depends on (a) the manufacturer designing the equipment for ease of repair and (b) the operating plant having on hand or on call an adequate source of spare parts and repair people with the required skills as well as easy access to the equipment needing repair.

By using MTBF and MTTR, we can determine the *availability* or *time-availability,* A, of an instrument or instrument system. This is the fraction of time that the instrument or system can be expected to be ready for use and to perform properly. "A" equals MTBF divided by the sum of MTBF and MTTR. The availability increases if MTBF increases or MTTR decreases.

Quite often, a manufacturer cannot provide data for MTBF and MTTR for calculating the availability, particularly for nonelectrical devices. However, it can be worthwhile to attempt an estimate of the availability or to judge the apparent quality of the devices. By taking reliability into account when choosing an instrument or planning a system, the plant will tend to have lower annual repair costs for faulty instruments. Also, plant downtime because of instrument failure and the resulting losses from nonproduction will be minimized. These factors may be influential in justifying a purchasing decision made in favor of equipment that will have the lowest cost over the life of the plant even if its initial cost is not the lowest.

11-2-2 Number of Loop Elements

Reliability is improved by reducing the number of links in the instrument chain. Section 11-1 discussed the overall accuracy of a specific measurement loop consisting of four instruments: sensor, transmitter, signal converter, and receiver. Each instrument has some risk of failure and contributes to the risk of loop failure. The loop risk can be reduced if we arrange the transmitter output signal to match the requirement for the receiver input; then we can eliminate the signal converter and its risk of failure. For this particular level loop, eliminating the converter would also lower the overall loop error from ±1.9% of span to ±1.15%, calculated by the algebraic method.

The principle of direct communications, or the lack thereof, and its effect on reliability are further exemplified in the children's game of "Telephone," which is played as follows. A group of children sit in a line. The first child whispers a message into the ear of his or her neighbor, and the message is thus relayed down the line. The point is to see how distorted the message has become when it gets to the last child. The more children there are in the line, the greater is the distortion—and the greater the laughter.

11-2-3 Redundancy

If an instrumentation failure in a plant could result in an unacceptable risk of physical danger or monetary loss, then *redundancy* may be called for. Redundancy means providing an alternate element to take over if the first element fails. Redundancy may be applied to any kind of equipment: sensors, controllers, computers, power supplies, fans, heat exchangers, entire systems, piping, and so on. A specific example of redundancy is the dual-channel data highway used in a distributed control system (DCS), as shown in Figure 8-5. If one channel fails, the other carries on while the failed channel is repaired.

For redundancy to be fully effective, each channel should operate totally independently of the other. This means that no single misoperation, such as incorrectly opening or closing one switch, and no single failure, such as a failure of a common power supply, could disable two channels. The failure of a common power supply is an example of a *common-mode failure*. But common-mode failure may also be caused if a single falling object disables two channels by damaging two redundant controllers, for example. To prevent such failure, the two channels should be protected by *separation*: that is, spacing the channels far apart from each other or providing adequate barriers between them.

Another way to increase plant reliability is by *diversity*. This calls for redundant channels to do the same thing but in different ways. The different channels are then unlikely to suffer all of the same kinds of failure. For example, a level measurement may be made by a displacement-type instrument and a head-type instrument, which measure level by different physical principles and are constructed differently.

A good design principle to follow in all plants is to separate the normal control function from a safety function. Figure 11-1(a) shows how normal control of tank level and the safety shutoff of a pump to prevent flooding can be accomplished. The level controller LC-101 modulates the flow so the tank does not normally flood. In case of high level from a severe upset, the high-level switch LSH-101 stops the pump and the tank inflow if the level becomes excessively high. The tank does not flood.

All parts of the scheme will probably work as they are supposed to whenever they are needed. However, the scheme has a weakness that can potentially cause trouble: Both the controller and the level switch depend on the single level transmitter and are, therefore, subject to common-mode failure.

Figure 11-1(b) shows a more reliable way to protect against flooding. Everything is the same as before except for the level switch LSH-103, which now senses the tank level directly. The control loop may fail, but the switch is not affected. On the other hand, if the switch fails, the control loop is not affected.

A less technical example of good reliability is that of a man holding up his pants by using a belt and suspenders. If one device fails, he expects the other to keep the pants from falling down. His safety system has redundancy, diversity, and separation.

Some processes have problems keeping sensors in good working order. The sensors tend to foul, for example, and become unreliable. Let's assume they give a false high reading. Redundancy may be applied by using three separate channels to actuate an alarm if any channel reading is high. The operator can then investigate and use his or her judgment about shutting down the process. But if any two channels read high, there is an automatic shutdown. The alarm is initiated by what is known as a *one-out-of-three* voting system and the automatic shutdown by a *two-out-of-three* system. The idea is that a single high reading could be an aberration and false and should not necessarily call for drastic protective action. But if that reading is confirmed by a second reading, then the high readings are considered valid. An even more conservative scheme uses a *two-out-of-four* system, which has four channels instead of three.

Another system measures disparities between two or more process instruments that should be giving the same reading. If a disparity becomes excessive, it actuates a disparity alarm even though a process problem may not be apparent. Measures to increase reliability may be applied to any process system of great potential hazard, such as the nuclear power industry.

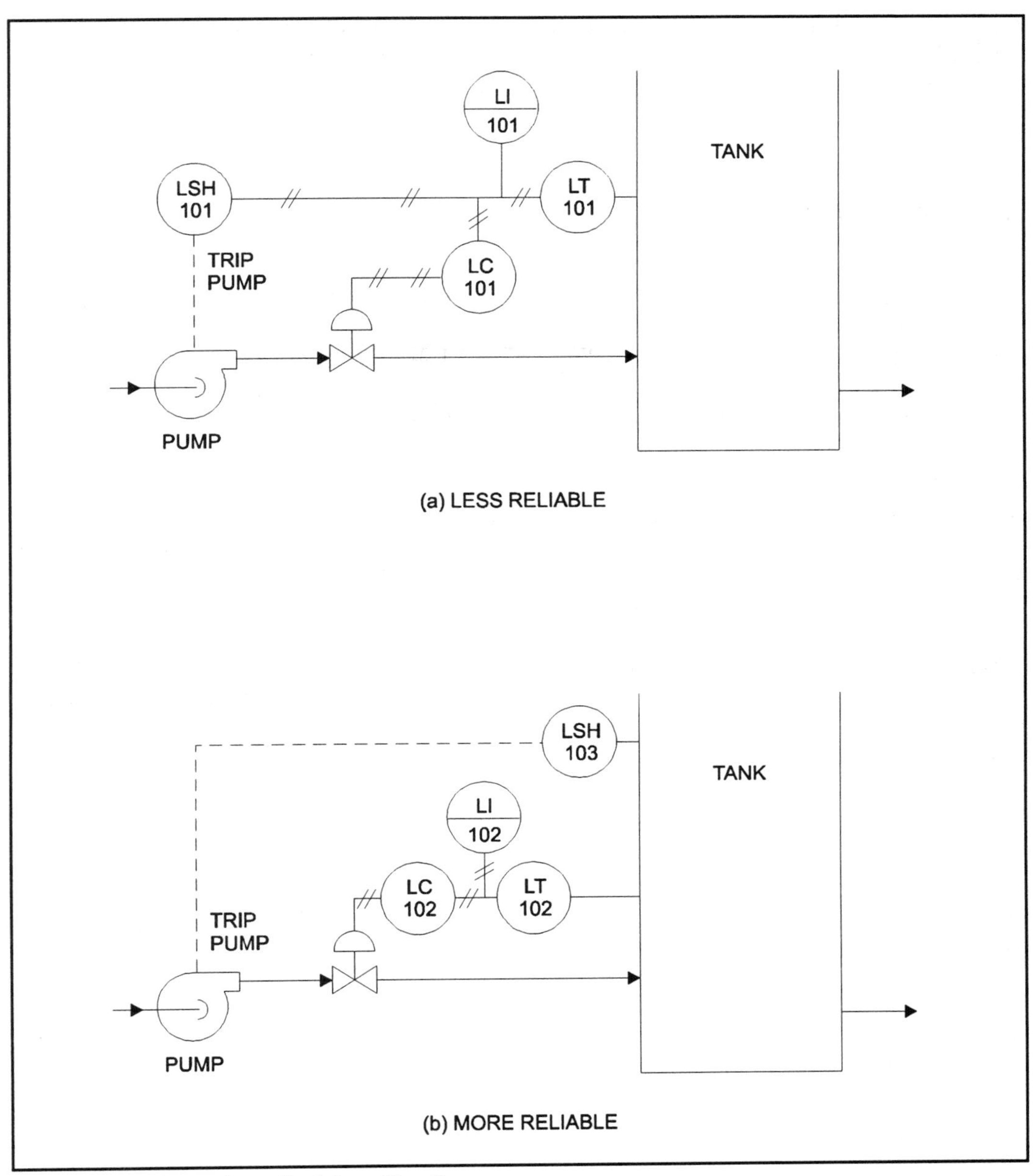

Figure 11-1. Preventing a Tank from Flooding

There is a movement in the world of electronics, including instruments, toward the development of so-called *fault-tolerant equipment*, which has internally redundant components or circuits. The effect is to enable the instrument or system involved to continue functioning properly even if some instrument part fails. This technique is used extensively in some DCSs and PLCs.

For important process systems, a *failure analysis* may be made. This is a detailed study of what may happen to the process if the various pieces of process equipment and instrument systems undergo all credible combinations of failure and mishap. The study may reveal a need for backup equipment, a change to a different type of equipment, a change of fail-safe action (see Section 6-1-4-4), or other changes. Or it may simply confirm the adequacy of the existing system design.

11-2-4 Trustworthiness

Instruments do not get bored, their attention does not wander, they do not listen to a ballgame, they do not stop to eat or go to the rest room, they do not come to work tired, they do not have emotional problems, and they do not abuse their bodies or their minds. They are always present and always attentive.

A process operation may be carefully and skillfully engineered to operate most efficiently or most safely in one particular manner. Operations people, sometimes at a low level, have been known to decide, without understanding the details and subtleties of the original design engineering, that the process should operate differently, and they change the control system accordingly. The process is thereby degraded. However, instrument systems can be arranged to have varying degrees of security to guard against tampering.

Barring malfunctions or tampering, for which various levels of protection can be provided, a properly designed instrument system can be counted on to do its job exactly the same way every single time. It is trustworthy in a way that no human being can be, and it contributes importantly to the reliability of the plant.

11-3 Human Factors

In days gone by, instrument boards were merely an assemblage of panels on which instruments were placed, generally with some reasonable degree of logic and common sense. This was done according to the individual case and without a systematic approach to improving the performance of the man/machine combination. Then followed centralized control, with its mushrooming amount of information for an operator to supervise. Later came nuclear power plants with their much publicized special emphasis on correct operation.

The process industries then became more involved in a discipline known by various names, including *human factors engineering*, *human engineering*, *ergonomics*, and *biotechnology*. The industries were able to draw on principles developed by the military, who use large quantities of technical equipment and manpower. When our military means business, their stakes are very high, so they have been active in developing the principles and practices that enable humans to make more effective use of themselves and their machines of defense and offense.

Human factors engineering has been applied to things as simple as a screwdriver, resulting in a screwdriver handle shaped as a ball instead of the traditional plain or fluted cylinder. The ball enables the person using the screwdriver to grip it more firmly, get more twist when turning the screw, and be more comfortable while doing this. However, even with the simple new design there are subtleties: What is the optimum size of the ball handle for most people? What surface finish shall be used for maximum grip with minimum discomfort?

The work of a control room operator involves many factors, described in the following sections.

11-3-1 Use of the Human Body

The use of the human body is covered by *anthropometry,* which is the study of the measurements of the human body, its ability to see and hear, its tendency to fatigue, and the like. This information is based on extensive research and testing. These body characteristics as well as the nature, frequency, and difficulty of the physical and mental tasks that the operator must perform, the position of his or her body, and his or her mobility requirements are important in designing instrument boards. They affect the dimensions and instrument layout of the board, even the appearance of the instruments.

Little things can affect an operator's performance over an eight-hour shift. An example for a video screen includes such questions as: how good is the picture resolution? how high is the screen placed? is it aimed properly? is it protected against glare?

Other anthropometric factors include the following:

1. *Physical Environment*. This includes items like room temperature, humidity, lighting, noise, and the amount of traffic in the room.
2. *Instrument Arrangement*. Principles for arranging instruments on a board for the operator's benefit include the following four rules of thumb.
   a. Instruments should be grouped logically to show their relationships to the process.
   b. Controllers or control stations and their associated readout instruments should be next to each other so far as practical.
   c. Similar groups of instruments should be arranged similarly, whether left and right or up and down, as illustrated in Figure 11-2.
   d. No more than five similar-appearing instruments should be in a line with uniform spacing unless there is color coding or other striking differentiation between neighboring instrument groups. The coding may be of panels or instrument bezels, for example. (The *bezel* is the rim that holds the glass or plastic covering on the face of an instrument.)

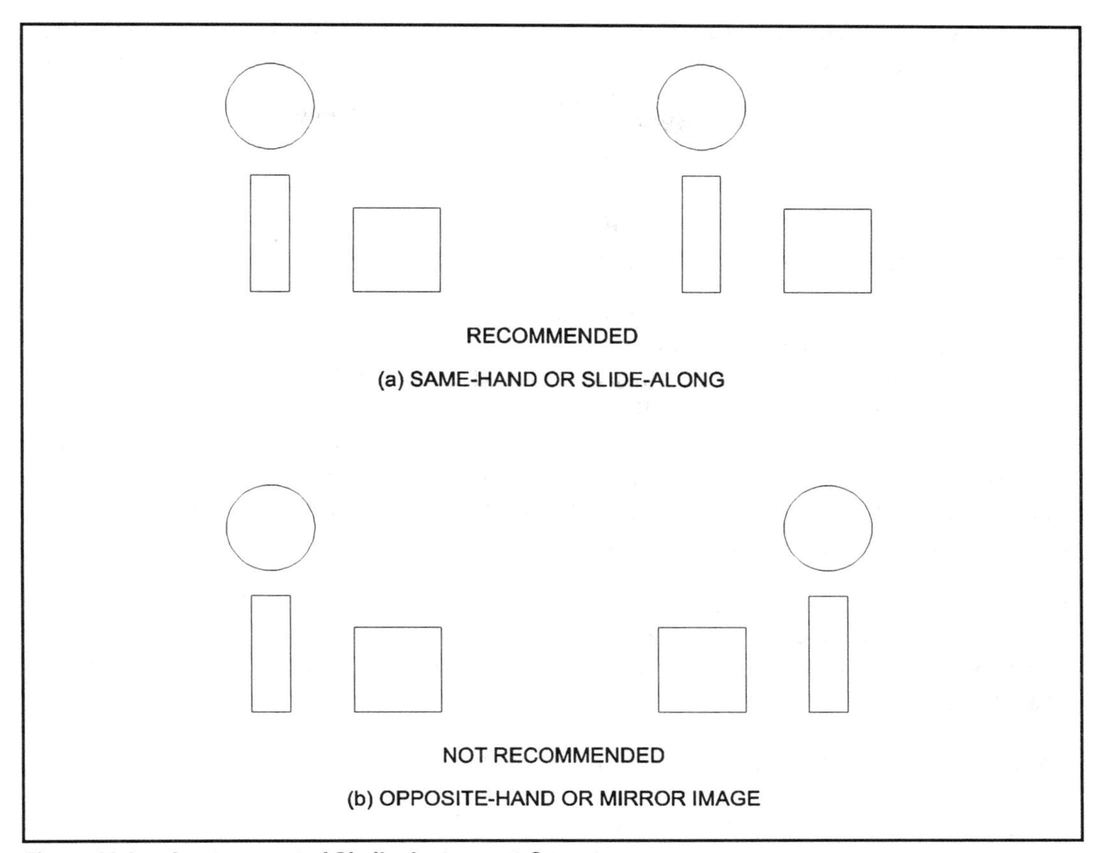

Figure 11-2. Arrangement of Similar Instrument Groups

11-3-2 Presentation of Information

In a large and complex plant, providing information by exception is a useful way to make life easier for an operator and is better for the plant. Information that is routinely presented to the operator should be limited to what he or she needs for routine tasks; more than that is a distraction. If trouble arises, then a second level of information should become available for the problem area. The hierarchy of video displays—overview, group, and detail—that can be provided in a DCS or PLC is helpful in permitting an operator to zoom in on what is important at any moment (see Section 8-3-2-6).

Red is our traditional color to signal trouble. We have red traffic lights and red signal flags that mean *Stop* or *Danger*. Red ink on the bottom line of a profit-and-loss statement for a business is bad news: it dramatically tells us that the business is losing money and is said to be "operating in the red."

An instrument board should never show any red light when everything works as it should and conditions are normal, although this is not universally practiced. For example, the color red should not be used for a pilot light to signal *Power On* as a normal condition. A different color should be used. If a red light ever does show up, it should have only one possible meaning—Trouble—and it should require a very prompt corrective response from the operator. In any emergency, he or she should not have to lose time sorting out whether the new red light is one of the abnormal red lights on the board or one of the normal red lights, which he or she will very likely find to be of no help at that time.

Another illustration of human factors engineering in presenting information is the use of different alarm sounds for different process groups, such as using a horn for one group of systems, a buzzer for a second group, and chimes for a third group.

11-4 Standardization of Hardware

When specifying instruments for purchase, it is worthwhile to consider to what extent the choice of kinds of instruments and their suppliers should be limited in order to standardize the hardware within the plant. In this context, the word *standardize* can mean the following things:

1. Where it is technically reasonable, handle similar situations in a plant in the same way. Where appropriate, measure various flow rates, for example, with the same kind of instruments. However, costs should be taken into account so that the kind of flow instrument used for small pipelines may end up different from that for large pipelines. Different fluids and different accuracy requirements may reasonably justify or mandate deviations from the standard.
2. So far as is practical, specify only a certain brand or brands of instruments. Keeping to a prescribed brand for a given plant may offer the following three advantages.
   a. The spare-parts inventory can be smaller because of the commonality of parts among instruments. This results in savings with respect to investment in inventory and storage space.
   b. Many instruments are interchangeable so a sudden need for a replacement is more apt to be filled by an existing instrument.
   c. The instrument installation people and service people work with already familiar devices. They have less need to refer to installation and service manuals, and they may work quicker and better.

The primary consideration must always remain to procure instruments that the applications truly require, not to adhere mindlessly to arbitrary purchasing rules.

11-5 Instrument Specifications

At some stage, the planning and design of instrument systems require purchasing, or other means of procuring, the instruments and auxiliary hardware that are needed to carry out the designs. When we go to buy something, we need to tell the seller what we want him or her to sell. The product description is normally stated on a document called a *specification, specification sheet, data sheet, requisition, bill of materials,* or otherwise. The document may refer to auxiliary documents—drawings, lists, codes, or other. If it does, the repetition of details on different documents should be avoided so far as practical. Repetition opens the door to errors and inconsistencies.

There may be sound technical or practical reasons for requesting certain nonstandard features on instruments. For example, the plant maintenance department may have found that a certain type of lubrication fitting for control valves can save the plant money over the years by facilitating the servicing of the valves. Therefore, a requirement for the control valve manufacturer to have these fittings installed at the factory could be justified.

Nevertheless, it is common to find overspecification of instruments on the part of the purchaser, that is, demanding more physical or functional features and details than are specifically needed. This can apply, for example, to an electrical connection for a power supply. Certainly, the instrument needs a connection, but must the purchaser demand that the connection have a 3/4-inch thread? The manufacturer's standard connection may be a 1/2-inch thread or some other conventional type. He or she may be able to furnish what the customer requests, but the order may then become special. There are many other ways of overspecifying, and all of them tend to cause one or more of the following undesirable and avoidable results:

1. Increased cost.
2. Lengthened delivery time.
3. More mistakes by the manufacturer.
4. Shutting out a good potential bidder who cannot or will not deviate from the bidder's standard design.

There are rules that, when intelligently applied, help to reduce overspecifying:

1. Specify every detail that is needed and that may cause trouble if it is not specified. For example, specifying simply that an instrument should be enclosed is not sufficient if a NEMA Type 4 enclosure (watertight and dust-proof) is required.
2. Do not specify anything that is not needed. Do not go into more detail than is necessary.

In short, the first thing to do is think. Then be willing to accept the manufacturer's standard product unless there is a specific reason not to.

12

Managerial Considerations

Suppose we are in charge of a production plant, and we believe the plant will operate better if it has a better process control system with new and expensive chemical analyzers and an elaborate sampling system. We are now considering whether or not we should make the change. We are not in business for our health, so we will use the old standby, money, as the criterion for our decision. We will not base the decision on vanity factors, such as "being on the cutting edge of progress." The hard question is: Will we gain more by the new system than we will lose?

Both benefits and costs of a proposed system are affected by the anticipated economic life of this system as compared to that of the existing system. Also to be considered are the effects of tax regulations, present and future, and other government policies relating to monetary inflation, interest rates, and even international trade. Governmental actions are not infrequently imponderables, being based as they are on human fallibility and passions and the often self-appointed right of government to manipulate the economic affairs of people and companies. These actions create a high degree of unpredictability in investment planning. However, the advantages of some considered investments are clear-cut enough to justify making a decision.

12-1 Benefits of the Proposed System

Regarding the benefits of a proposed new system, the following questions arise:

1. *Greater Stability.* Will it provide better control of the plant operation and reduce the number and magnitude of harmful process upsets?
2. *Improved Quality.* Will the system improve product quality, thereby reducing or eliminating the cost of reruns or wastage? Will it provide a higher-grade product for which a premium price may be charged?
3. *Improved Production.* Will it increase production and sales?
4. *Higher Efficiency.* Will it raise plant efficiency, resulting in higher production or lower requirements for feed materials and energy?
5. *Improved Reliability.* Will it be more reliable in keeping the plant operating and avoiding the costs of downtime and repairs?

6. *Greater Safety.* Will it be safer and cleaner? Will it help to avoid penalties for violating governmental regulations? Will the system lower or increase insurance costs?

12-2 Costs of the Proposed System

The other side of the evaluation of new systems concerns its costs, with the following questions:

1. *Cost to Buy and Install.* How much will the system cost to buy and install?
2. *Plant Downtime.* Will the system require plant downtime for installation? What will this cost?
3. *Manufacturer's Field Service.* Will the system initially require field service by the manufacturer? How much will this cost?
4. *Future Service.* What will the increase be for the future cost of the servicing, repair, and the associated plant downtimes for the new system as compared to the future cost, if any, for the present system?
5. *Incidental Plant Changes.* What incidental plant changes will be required for the instrument system, and what will they cost? These changes may include added piping, expansion of the power supply system, structural modifications, additional or altered instrument supports and boards, and so on.
6. *Future Operating Cost.* Will there be a significant change in operating cost to use the new system?
7. *Availability of Investment Capital.* Where will the money come from to pay for the new system? What will the money cost?

An additional factor that may have a bearing on the costs of a new system involves with who will do the work of instrument system design, purchasing, and installation, which raises the following questions:

1. *Design.* Who will do the detailed design for the contemplated instrument system? If a new process system design is purchased, will the process design company design the instrument system? Or will the instrument system be designed by the headquarters staff of the operating company? an engineering contractor? a full-line instrument manufacturer?
2. *Purchasing.* Will the designer of the instrument system supply the system? If the instrument system is both designed and purchased by the company that designed the system, then there is a single source that is responsible for the process and instrument package. If an operating problem arises, that company supposedly cannot point the finger of responsibility at someone else.

Will our operating company purchase the required instrumentation? The responsibility for purchasing may carry with it the responsibility for coordinating with the company that designs the instrument system and for performing any inspection work that may be necessary. Does our company have the manpower and skill to handle this additional work?

3. *Installation.* Will the instrumentation be installed by our operating company? by the instrument supplier? by outside contractors, with construction management performed by the operating company? by a general contractor who will subcontract and manage as required?
4. *Labor Relations.* Are we compelled to use union labor?
5. *Coordination.* What is the most effective arrangement overall for coordinating the different phases of the work?

12-3 Evaluating the Proposal

Sections 12-1 and 12-2 listed a number of the benefits and costs of our proposed new control system. If the proposed system can show a net return that is significantly greater than the return for the existing system, then the proposed system would appear to be the better economic alternative. However, there may be still another alternative: Is this the best use of our money? Can we and are we willing to invest that same money for a greater return in some way not connected with our present business? Also, is the plant investment worth the risk (which depends on many uncertainties)?

The soundness of our decision is vitally dependent on how accurate our estimates of benefits and costs turn out to be. When considering the economic climate of the future, a good deal of educated guesswork, crystal-ball gazing, and finger crossing may be required. Yet, despite the uncertainties, a conscientious estimate is better than none at all and, for those of us who are not certified extrasensory perceptionists, better than working by hunch. A poor decision may result in serious consequences, such as the following:

1. *High Cost-to-Benefit Ratio.* The costs may be unnecessarily high compared to the benefits received. This may apply to the capital investment, the annual costs, or both.
2. *Poor Performance.* The plant performance may be inferior, again resulting in financial consequences.
3. *Bad Side Effects.* There may be adverse effects on the design or operation of equipment or the process in other parts of the plant.
4. *Possible Lock-In.* Assume that a poorly chosen system was purchased and must be paid for regardless of whether the system is kept in service or gotten rid of at once. Because the purchase cannot be undone, the investment is known as a *sunk cost.* All things considered, it may then be less costly to live with the poor choice from here on than to discard it and replace it. In such a case, the poor decision is irrevocable in practical terms.

APPENDIX A

TYPES OF INSTRUMENTATION DOCUMENTS

The systematic development of instrument systems for a specific plant project requires that certain documents be prepared. Most of these documents are prepared or initiated by instrument engineers, while some are developed by other disciplines and are also used by the instrument engineers.

These documents enable the instrument engineers to do the following:

1. List their work as it progresses.
2. Communicate with their co-workers, their client, manufacturers and their representatives, and construction people.
3. Provide reference documents for themselves, the field people, the plant engineers, the operators, and the service people.

Different organizations doing the same kind of work operate more or less differently from each other, and their paperwork may be somewhat different. However, in one way or another, they perform essentially the same functions.

For repetitive details of design, organizations normally simplify their work by making use of standards that they themselves produced as well as national standards published by ISA and other engineering societies. A *standard* is a model or set of rules that defines the requirements for a design, performance, or procedure for a specific purpose.

The use of standards speeds up the work because the answers to many questions concerning the subject matter of the standards are already there. At the same time, their use generally improves the quality of the work because each standard normally receives critical review before becoming official. In the case of national standards, such as those of ISA, the standards represent a consensus of good industry practice.

The engineering documents that instrument engineers produce are typically those described in the following sections.

Plant Process Diagrams

Process Flow Diagrams

These are drawings to define the process. They are prepared by the process engineer, the person who designs the process, in cooperation with the work of the instrument engineer regarding matters of measurement or control. The process is the sole reason for having the plant; it is the beginning point of the project and the basis for requirements for land, energy, types and sizes of equipment, and controls. It is also valuable after the plant is operating.

The process flow drawing shows the major process equipment and streams as well as the fluid composition, pressure, temperature, flow, and other essential operating conditions at different places and for different cases of operation. It shows the barest minimum of instruments needed to indicate which process variables need to be controlled or manipulated.

Piping-and-Instrument Diagrams (P&IDs)

The P&ID is an expanded version of the process flow diagram but without the detailed data on operating conditions. It shows all the process piping, including valves and other devices; all the equipment for the process; and utility systems for the plant: air, steam, fuel, and others. All these are handled by the process engineer.

The P&ID also includes all the instruments that the process requires, including minor instruments, such as thermometers and pressure gages. Incidental instruments, such as air supply regulators for individual instruments, are generally not shown. The instruments are added to the drawing at the direction of the instrument engineer.

If a process subsystem—for example, a water-treating system—is purchased as a package, then the P&ID may represent it with a rectangle that shows no process equipment or instruments. Instead, the box carries a reference to a manufacturer's drawing that shows all these details. An alternative is to draw the package system details on the P&ID in order to complete the picture there.

The P&ID schematically tells the piping designer where to place instrument connections and straight runs of piping when they are needed for measurement. The diagram is also used by many disciplines and by the operating company as an overall functional picture of the plant for use in discussions and for reference by the operators, maintenance people, and so on.

Different companies may refer to the P&ID diagrams by the following more or less equivalent titles: *engineering flow diagram*, *mechanical flow diagram*, and others.

Instrument Criteria

Instrument Application Criteria

The instrument engineer for a particular project prepares a document that states the guidelines that should generally be followed for selecting instruments. In effect, the guidelines become project standards to promote uniformity and quality. They include items such as:

1. Standard signal ranges.
2. The types of sensors that should generally be used and the limitations on their use.
3. The variables that are to be put into the plant computer.
4. The types of annunciators and other alarms, their sequences, and their color schemes.
5. The size and color of the instrument cases to be mounted in the control room.
6. The preferred brands of instruments. These may be based on the instrument engineer's experience and judgment but may also reflect the operating company's preferences.
7. Many other points of instrument application philosophy and practice.

Instrument Installation Criteria

This document is prepared by the instrument engineer for a specific project in order to establish the guidelines to be generally followed for installing instruments. The document covers such requirements as the following:

1. *Root valves and instrument valves*: their types and sizes.
2. *Sensing lines*: their sizes, bleed connections, test connections, and general arrangement.
3. *Tubing materials, and the size and type of tube fittings.*
4. *Installation tolerances.*

Loop Diagrams

A loop diagram shows the detailed interconnections of the instruments of a loop and the instrument connections to power supplies. The diagram may cover system loops that are electric, pneumatic, hydraulic, fiber-optic, or a combination thereof.

Logic Control Diagrams

Logic control diagrams define the on-off functions of instrument systems, as covered in Chapter 7.

Instrument Specifications

These documents state the descriptions and requirements for instruments, instrument systems, and auxiliary equipment and services—such as start-up assistance by the instrument manufacturer—that are to be purchased. ISA Standard S20, *Specification Forms for Process Measurement and Control Instruments, Primary Elements, and Control Valves*, provides different fill-in specification forms for process measurement and control instruments. These forms are also available as computer software.

Computer Documents

Plant computers need supporting documents that the instrument engineer prepares, which include the following:

1. *Computer Input/Output (I/O) List*. This list includes the following three items.
    a. The input signals and for each its source, computer address, type—analog, binary, or digital—as well as its range and engineering units.
    b. The functions that the signals are used for, for example, signal conversion, calculation, routine or abnormal display, or alarm.
    c. The output signals and for each its computer address, destination, type—analog, binary, or digital—as well as its range and engineering units.
2. *Computer Logic Diagrams*. Plant computers can be purchased with all their programming done by the supplier(s) according to operating requirements and computer logic diagrams outlined by the instrument engineer. The programming defines the way that the computers perform their control and information functions.
3. *Computer Display Diagrams*. The computer video displays may be of lists, statements, pictures, or a combination of these. Information for these displays is developed by the instrument engineer and given to the supplier for programming.
4. *Wiring Diagrams*. The instrument engineer prepares wiring diagrams or lists to enable the electrical designer to plan the wiring between the computer systems, the process instruments, and the power supplies. The diagrams identify the computer system's electrical terminals and show the devices to which they are wired.

Annunciator Diagrams

Annunciator diagrams show the annunciator as well as an arrangement and list of windows, the annunciator's wiring points, and the sources of the annunciator's input signals. If the annunciator has electrical outputs, these and their destinations are also listed.

Construction Diagrams

Instrument Location Diagrams

These indicate the locations and elevations of instruments relative to the plant coordinates.

Console and Instrument Board Diagrams

These show the outlines of consoles, instrument boards, how the two are sectioned, and how the instruments and other devices are laid out.

Instrument Installation Diagrams

These show the piping or tubing for connecting instruments to the process, to other instruments, and to pneumatic power supplies. They also describe methods for supporting instruments and give other details of installation. They may also include instructions for special calibration to be performed in the field for individual instruments.

Instrument Heat-tracing Diagrams

These list the instruments and sensing lines that require heat tracing or other types of weatherizing and the desired control temperature for each application. They may describe how the tracing should be applied.

Purchase Requisitions

The instrument engineer prepares a purchase requisition, which is a written request for the purchase of instrumentation or services. The requisition contains a reference to an attached specification for the items and to other descriptive material, if needed. A requisition is usually the basis for the invitations to bid that are sent to suppliers. After the instrument engineer reviews the bids, he or she may revise the specification and the requisition. The purchasing department then issues a purchase order to the successful bidder.

System Description

The system description is a description of the design and operation of the plant process, prepared by the process engineer, with additional material written by the instrument engineer and others. The instrument engineer's contribution is to discuss the instruments and how they are used to control the process.

The system description is a valuable reference document for everyone who gets involved in the working of the plant. It describes the plant as it was when built, and it should be kept up to date if changes to the plant are made.

Instrument Index

The instrument index is a tabulation that the instrument engineer prepares to keep track of all the instruments and their related documents. In its pure form, the index provides no direct information about the instruments but is merely a directory that points out where the information may be found. Substantively it says very little, yet it is a major work document for the instrument engineer and others during the design, construction, and operation phases of the plant.

The initiation, expansion, use, and maintenance of the index begin with the preparation of the piping-and-instrument diagram (P&ID) near the start of the project for work on a plant and continue throughout the entire project. Listing an instrument in the index before any engineering work is done on it helps to ensure that it is not later overlooked and that identification numbers are not duplicated.

For a large project, the index may consist of hundreds of pages of fifteen-inch-wide computer paper. A computerized index can readily obtain lists of instruments that still need to be purchased, can count instruments of a given group, and so on.

The tabulation in the index may include column headings such as those shown in the following list. For each instrument listed in the index information is supplied under each heading as appropriate and as the information becomes available. The documents described earlier in this appendix are included in the following list:

- Instrument identification number, for example, FE-1203. If instruments are provided and identified by a system-package manufacturer, then the manufacturer's instrument identification numbers may be listed.

- Instrument service, for example, "product to storage."

- P&ID number.

- Specification number.

- Purchase order number, if different from the specification number.

- Loop diagram number.
- Logic control diagram number.
- Computer input/output (I/O) list number.
- Computer display diagram number.
- Wiring diagram number.
- Electrical schematic diagram number. This diagram is generally made by an electrical designer on the basis of the logic control diagrams prepared by an instrument engineer.
- Annunciator drawing number.
- The number of the process piping or equipment drawing that shows the sensor connection. These drawings are generally made by a piping designer or an equipment designer.
- Instrument location diagram number.
- Instrument installation drawing number.
- Heat-tracing list number.
- Instrument board drawing number.
- The numbers of the manufacturer's documents, such as outline drawing, wiring diagram, installation manual, service manual.
- Remarks.

APPENDIX B

GAS VERSUS VAPOR

The words *gas* and *vapor* are not synonymous, although they are sometimes used loosely and interchangeably. This appendix describes the difference between them.

When a pot of water is boiled, the liquid evaporates and becomes steam. The steam is an invisible vapor. If we see a mist or fog rising from the water in the pot, we still do not see the steam. What we see are fine droplets of water that condensed when the steam met cool air.

Steam can be condensed back to the liquid state by either cooling or compressing it. If we put the vapor in a heater that raises the temperature above 705.40°F (374.11°C), the steam cannot condense, no matter how much it is compressed. This temperature is known as the *critical temperature* of water. Above the critical temperature, water is a gas. Below that temperature, it is a vapor or, if it gets cold enough, it is a liquid; and if it gets still colder, it becomes ice.

Every substance has its own critical temperature above which it is impossible to liquefy the substance and above which the substance is a gas. Examples of these temperatures are as follows:

Substance	Critical Temperature	
Nitrogen	–233°F	(–147°C)
Oxygen	–182°F	(–119°C)
Methane	–117°F	(–83°C)
Propane	206°F	(97°C)
Ammonia	270°F	(132°C)
Chlorine	291°F	(144°C)
Ethyl alcohol	470°F	(243°C)

The pure air that we breathe consists almost entirely of nitrogen and oxygen, which are gases so far as we are concerned because they are normally far above their critical temperatures. Propane, the so-called *LP gas*, which is sold in bottles for use in portable stoves, actually exists as a vapor at room temperature. Gasoline fumes are vapors. Alcoholic drinks release ethyl alcohol vapor into the air.

INDEX